PRECISION-GUIDED MUNITIONS AND HUMAN SUFFERING IN WAR

Military and Defence Ethics

Series Editors

There is an urgent and growing need for all those involved in matters of national defence – from policy makers to armaments manufacturers to members of the armed forces – to behave, and to be seen to behave, ethically. The ethical dimensions of making decisions and taking action in the defence arena are the subject of intense and ongoing media interest and public scrutiny. It is vital that all those involved be given the benefit of the finest possible advice and support. Such advice is best sought from those who have great practical experience or theoretical wisdom (or both) in their particular field and publication of their work in this series will ensure that it is readily accessible to all who need it.

Also in the series

Protecting Civilians During Violent Conflict
Theoretical and Practical Issues for the 21st Century
Edited by David W. Lovell and Igor Primoratz
ISBN 978-1-4094-3125-1

Kantian Thinking about Military Ethics
J. Carl Ficarrotta
ISBN: 978-0-7546-7992-9

Empowering Our Military Conscience
Roger Wertheimer
ISBN: 978-0-7546-7894-6

Precision-guided Munitions and Human Suffering in War

JAMES E. HICKEY
U.S. Naval War College, Newport, USA

Routledge
Taylor & Francis Group
LONDON AND NEW YORK

First published 2012 by Ashgate Publishing

Published 2016 by Routledge
2 Park Square, Milton Park, Abingdon, Oxon OX14 4RN
711 Third Avenue, New York, NY 10017, USA

First issued in paperback 2017

Routledge is an imprint of the Taylor & Francis Group, an informa business

British Library Cataloguing in Publication Data
Hickey, James E.
Precision-guided munitions and human suffering in war. –
(Military and defence ethics)
1. Precision guided munitions--History. 2. Military
weapons--Technological innovations. 3. War and society.
4. Military ethics. 5. Casualty aversion (Military science)
6. Military art and science--History--20th century.
I. Title II. Series
355'.02'09045–dc23

Library of Congress Cataloging-in-Publication Data
Hickey, James E., Ph. D.
Precision-guided munitions and human suffering in war / by James E. Hickey.
p. cm. -- (Military and defence ethics)
Includes bibliographical references and index.
ISBN 978-1-4094-2951-7 (hbk)
1. Precision guided munitions. 2. Civilian war casualties. 3.
Air warfare--Decision making. 4. Bombing, Aerial--Moral and ethical
aspects. 5. Military ethics. 6. Just war doctrine. 7. Precision guided
munitions--United States--History. I. Title.
UF510.H53 2012
172'.42–dc23

2012004875

ISBN 13: 978-1-138-11569-9 (pbk)
ISBN 13: 978-1-4094-2951-7 (hbk)

Contents

List of Tables

Acknowledgements

Writing the acknowledgement page for any work is fraught with the danger of leaving someone out who deserves special mention. If I did so, the fault rests solely with me and I apologize. I would first like to thank Ashgate Publishing and, in particular, Margaret Younger, who has shepherded me through the process. My thanks also to James Connelly, the series editor, for his active support when it was needed most. To Martin Cook and George Lucas, thank you for graciously pointing me in the right direction in the first place. To the countless others, too numerous to name individually, who kept me focused by asking how the project was progessing—you motivated me more than you will ever know, particularly those of you who were on board from the very beginning.

For the last 15 years, I have been blessed with the opportunity to teach at two world-class professional military education institutions. To my colleagues at the United Kingdom's Joint Services Command and Staff College, and at the U.S. Naval War College in Newport, I offer my sincere gratitude for being permitted to teach among dedicated professionals. You have helped, in so many ways, to make me a better strategic thinker.

Next, I turn to my family. To my parents, John and Barbara, who encouraged me to keep pushing forward during this long journey. To my daughters, Erin, Shawn, and Caitlin, who often looked on with bemusement as research materials surrounded me at the dining room table. Finally, to Rae, who quite simply has always been there for more than 30 years. Without your support and encouragement, this would not be complete. Thank you.

Abbreviations and Acronyms

AAA	Anti-Aircraft Artillery
ACTORD	Activation Orders (NATO)
AFB	Air Force Base
AGM	Air-to-Ground Munition
AIRSOUTH	Allied Air Forces Southern Europe (NATO)
AOCC	Air Operations Control Center
ARVN	Army of the Republic of Vietnam
AZON	Azimuth Only
BBC	British Broadcast Corporation
BCE	Before the Common Era
BSA	Bosnian Serb Army
C3	Command, Control, Communications
CALCM	Conventional Air-Launched Cruise Missile
CAOC	Combined Air Operations Center
CAS	Close Air Support
CAT	Computerized Axial Tomography
CBU	Cluster Bomb Unit
CE	Common Era
CENTAF	U.S. Air Force, Central Command
CENTCOM	U.S. Central Command
CEP	Circular Error Probable
CIA	Central Intelligence Agency
CINC	Commander-in-Chief
CINCCENT	Commander-in-Chief, U.S. Central Command
CINCPAC	Commander-in-Chief, U.S. Pacific Command
CINCSOUTH	Commander-in-Chief, Allied Forces Southern Europe
CJCS	Chairman, Joint Chiefs of Staff
CNN	Cable News Network
COCOM	Combatant Command
CSCE	Commission on Security and Cooperation in Europe
DMPI	Desired Mean Point of Impact
DOD	Department of Defense
DSMAC	Digital Scene-Matching Area Correlation
EC	European Community
EO	Electro-Optical
EU	European Union
FLIR	Forward Looking Infrared

FON	Freedom of Navigation
FRY	Federal Republic of Yugoslavia
GAO	Government Accounting Office
GBU	Guided Bomb Unit
GCC	Gulf Cooperation Council
GDP	Gross Domestic Product
GOSP	Gas-Oil Separation Platform
GPS	Global Positioning System
GWAPS	Gulf War Air Power Survey
HOBO	Homing Bomb
IADS	Integrated Air Defense System
ICBM	Intercontinental Ballistic Missile
IFOR	Implementation Force
JCS	Joint Chiefs of Staff
JDAM	Joint Direct Attack Munition
JFACC	Joint Force Air Component Commander
JNA	Yugoslavian National Army
KDP	Kurdistan Democratic Party
KFOR	Kosovo Force (NATO)
KLA	Kosovo Liberation Army
KTO	Kuwait Theater of Operations
LDK	Democratic League of Kosovo
LGB	Laser-Guided Bomb
LOAC	Law of Armed Conflict
LOC	Line of Communication
MACV	Military Assistance Command, Vietnam
MC	Military Committee (NATO)
MRI	Magnetic Resonance Imaging
MTR	Military Technical Revolution
NAC	North Atlantic Council (NATO)
NATO	North Atlantic Treaty Organization
NSC	National Security Council
NSD	National Security Directive
NSSM	National Security Study Memorandum
NVA	North Vietnamese Army
OED	Oxford English Dictionary
OPEC	Organization of Petroleum Exporting Countries
OPLAN	Operation Plan
OSD	Office of the Secretary of Defense
OT&E	Operational Test and Evaluation
PGM	Precision-Guided Munition
POL	Petroleum, Oil, Lubricants
POW	Prisoner of War
PUK	Patriotic Union of Kurdistan

QDV	Quick Decisive Victory
RAZON	Range and Azimuth Only
RDF	Rapid Deployment Force
RMA	Revolution in Military Affairs
ROE	Rules of Engagement
SACEUR	Supreme Allied Commander, Europe (NATO)
SAM	Surface-to-Air Missile
SDB	Small Diameter Bomb
SECDEF	Secretary of Defense
SFOR	Stabilization Force (NATO)
SLAM	Stand-off Land Attack Missile
TAC	Tactical Air Command
TARZON	"Tall Boy" Range and Azimuth Only
TERCOM	Terrain Contour Matching
TFW	Tactical Fighter Wing
TLAM	Tomahawk Land Attack Missile
UAE	United Arab Emirates
UN	United Nations
UNHCR	United Nations High Commission on Refugees
UNMIK	United Nations
UNPF	United Nations Peace Forces
UNPROFOR	United Nations Protection Force
UNSCR	United Nations Security Council Resolution
USA	U.S. Army
USAF	U.S. Air Force
USMC	U.S. Marine Corps
USN	U.S. Navy
VB	Vertical Bomb
VC	Viet Cong
WMD	Weapons of Mass Destruction

To Rae

Introduction

Throughout history, humans have demonstrated a proclivity for using violence against one another as a means to achieve an end, means enabled, in many respects, by the technologies available at the time. For much of that history, advancing technology, the "systematic application of science or other organized knowledge to practical tasks," has been a primary enabler of ever-*increasing* levels of violence and attendant human suffering.[1] The development, in the twentieth century, of thermo-nuclear tipped intercontinental ballistic missiles was perhaps the ultimate expression of the evolutionary view of advancing technology as the harbinger of exponentially increasing lethality.

But strategic culture, ways of warfare, and, indeed, technology itself are all social constructs. This means that given the same set of circumstances and the same available technologies, societies will often seek far different solutions, in any number of fields, using far different means depending on their cultural perspective. Simply stated, a society's cultural focus, particularly on technology, will influence the types of artifacts it produces. Those artifacts subsequently influence decision-makers and policy development within that society, in a process that continually cycles and renews itself.

Journalist and author Robert Kaplan finds a nation's way of waging warfare particularly insightful. He writes, "Military campaigns, because they are fights for the sheer survival of nations and cultures, offer the most telling insights about the values, technologies, social relations, and intellectual life of historical periods."[2] In the last 40 years, the United States, for example, seems to have transitioned between two of these "historical periods."

The earlier "American Way of War," as described by Russell Weigley in his book of the same name, was characterized by massed forces seeking to annihilate the enemy on the battlefield, as evidenced in World War I, World War II, and Korea. Such a way of warfare tolerated, even justified, the bombing of civilians in an effort to end the war.

Contrast that approach with the reality of the "New American Way of War" described by Thomas Mahnken, Reuben Brigety II, and Eliot Cohen, among others. As observed in operations such as Bosnia, Kosovo and, more recently, Libya, massed forces have been replaced by carefully targeted "smart bombs" or cruise missiles. War now results in fewer casualties and a much greater emphasis

1 John Kenneth Galbraith's definition. Carl Mitcham, *Thinking Through Technology: The Path between Engineering and Philosophy* (Chicago: University of Chicago Press, 1994), 153.

2 2 Robert D. Kaplan, "Four-Star Generalists." *The Atlantic Monthly* 284.4 (October 1999): 18.

has been placed, by both national civilian and military leaders, on keeping civilian losses/collateral damage to a minimum.

Thus, advancing technology is now touted as being able to provide less lethal alternatives. The development of these precision-guided munitions (PGMs), with their high degree of discrimination and accuracy, enables the armed forces that possess them to fight wars with ever-*decreasing* levels of violence and suffering. However, caution is certainly in order. This is not the first time in history that people hoped for such an outcome. In the seventeenth century, similar hopes for both gunpowder and artillery went unrealized.

Nonetheless, the press is replete with examples of the fascination with such "smart" weapons. Pictures of Tomahawk Land-attack Missiles (TLAM) launched from U.S. Navy ships in the Arabian Gulf and film clips of cockpit footage showing precise surgical strikes into enemy territory are seen on televisions in households around the world. Indeed some, such as William Arkin, have argued that, "unsparing news media focus ... [resulted in] a hyperdiscriminate approach" to targeting during Operation DESERT STORM.[3] One military journal article, "Air War over Iraq Proves Need for Smaller, More Precise Ordnance," described operations "that require high precision but minimize civilian casualties and collateral damage."[4] These are compelling words. But does anyone fully understand the impact that the use of such weapons is having on human suffering?

For even with this apparently less lethal form of warfare available, the political decisions made by leaders taking a nation to war (or not) and in determining a nation's acceptable conduct in war, still have ethical implications and consequences. This realization presents two important questions. First, how should these decisions be evaluated? In other words, against which, if any, ethical framework should a given selected course of action be measured? Is it sufficient simply to evaluate the perceived correctness of a chosen course of action, a more Kantian approach, or is it better to look at the outcome, seeking the "best for the most," the utilitarian approach? Alternatively, while a fully developed ethical system should be complete, comprehensive, non-contradictory, and livable, should the study seek to cross ethical and philosophical lines to derive a blended solution? Secondly, why is such an inquiry important?

The second question is perhaps the easier to answer. The inquiry is important because, despite humanity's propensity towards the use of violence to achieve various ends, most societies appear uncomfortable with an "anything goes" attitude in its application. Most people, paradoxically, while apparently believing in the use of violence as a legitimate tool of statecraft are, at the same time, uncomfortable with its unbridled application—actively seeking to limit the incidence of, and conduct in, war. In reality, the *realpolitik* conducted on the national scale often

3 William Arkin, "Baghdad: The Urban Sanctuary in Desert Storm?" *Airpower Journal* XI.1 (Spring 1997): 8.

4 Sandra I. Erwin, "Washington Report: Air War Over Iraq Proves Need for Smaller, More Precise Ordnance," *The Hook* 33.2 (Summer 2005): 68.

involves actions that many find abhorrent on a personal level. By understanding the underlying rationale for how nations today deal with these two critical decisions, it may be possible to reduce human suffering in future conflict. This study seeks to gain such an insight.

When thinking about the resort to violence, ideas and thinking have ranged from ideologies of pacifism at one extreme, Gandhi being one well-known adherent, to rationales for "warmongering," or "amoral realism," more Machiavellian approaches at the other. An examination of the literature, however, reveals that most cultures and societies have sought a pragmatic middle-ground position, what Mark Amstutz calls the ideology of "principled realism." The Western ethical construct of the Just War Tradition is not the only touchstone for a pragmatic approach. Indeed a thorough review of numerous cultural, religious, and secular traditions demonstrates that most mitigation efforts rest upon *jus ad bellum* (when it is morally correct to go to war), *jus in bello* (what actions are morally permissible in war), and *jus post bellum* (correct actions after the "just war" is over)-like concerns.

Interestingly, even traditions seen by outsiders as almost exclusively pacifistic sometimes contain *jus ad bellum* debates. For example, "a vocal Buddhist minority in Sri Lanka ... consistently advocated a military rather than negotiated solution to the island's long-standing ethnic conflict."[5] Similarly, Confucius, while "rejecting the principles of pragmatism which people typically use to justify war" mentions military expeditions going forth to battle.[6]

A number of authors have written on the Just War Tradition in a modern context.[7] Others have written about the impact of advanced precision-guided munitions on the conduct of war in the last few decades. These authors have focused on the impact, primarily from a tactical and operational sense, such weapons have made by more effectively engaging and striking targets deemed critical to the war effort.[8] Still others tout the benefits that standoff precision weapons have on preserving the lives of friendly personnel. Finally, some authors have explored the difficult plight of defeated populations in the aftermath of modern war, populations often struggling without the critical infrastructure deliberately targeted during the war to achieve desired political objectives.[9]

5 Elizabeth J. Harris, "Buddhism and the Justification of War: A Case Study from Sri Lanka," in *Just War in Comparative Perspective*, ed. Paul Robinson (Burlington, VT: Ashgate, 2003), 93.

6 Thomas M. Kane, "Inauspicious Tools: Chinese Thought on the Morality of Warfare," in *Just War in Comparative Perspective*, ed. Paul Robinson (Burlington, VT: Ashgate, 2003), 141.

7 James Turner Johnson and Michael Walzer are prominent here. See Bibliography.

8 Colonel John Warden, Thomas Mahnken, Reuben Brigety, among others.

9 William Arkin has done much in this area, as has Alexander Downes. See also Human Rights Watch.

What is missing, in all of these more narrowly focused efforts, is a comprehensive look at the wider impact on humanity of modern conflict waged in large measure, at least on the part of the U.S., by precision-enabled means. This study seeks to do just that, to look longitudinally across the individual phases of the conflict to gain a greater understanding of the *overall* impact the use of such weapons has on human suffering. Therefore, the study adopts a multi-disciplinary approach to answer the question: Do PGMs mitigate suffering in war, and have these weapons changed the way decisions regarding war and peace are made?

The approach is partly historical, in the wider sense of "the history of ideas," and by the use of case studies spanning the period from the first large-scale operational use of PGMs in combat in Southeast Asia in 1967 to operations in Kosovo in 1999. The study also engages a number of critics of technology, primarily Jacques Ellul, examining the concerns these philosophers raise to see if they are valid in society today. Finally, while answering the question posed would be interesting as a study in itself, having transformational technology influences key decision-makers. As a result, the study also examines the policy implications of PGMs.

In order to investigate this over-arching "umbrella" question fully, the study, while firmly grounded in the Just War Tradition, engages a number of philosophical and ethical traditions to achieve a greater understanding of just war's intellectual underpinnings. Therefore, elements of the utilitarian tradition, primarily the hedonistic (pleasure) form of Jeremy Bentham and the eudaimonistic (happiness) form of John Stuart Mill, are integrated with elements of Kant's ethics of duty.[10] These three philosophers did not invent the concerns they wrote about; in fact, each explores a component part of Aquinas' earlier approach, highlighting and emphasizing it in a new systematic way that informs the study. Such an admixture is important because, while it is true that a utilitarian approach seeks to ensure the greatest good for the greatest number of people, humankind must also remember its duty to allow the least among us the opportunity to flourish.

Similarly, elements of natural law and Aristotle's ethics of virtue are present in the analysis. While certainly not perfect in every application, for over two hundred years America has sought to maintain a moral high ground in its national and international dealings. Such sentiments are embodied in its founding documents. The Declaration of Independence states, "we hold these truths to be self evident, that all men are created equal, that they are endowed by their Creator with certain unalienable Rights, that among these are Life, Liberty and the pursuit of Happiness." Elements of both natural law/rights and a more modern articulation of the Greek concept of *eudaimonia* or "flourishing" are clearly present.

What emerges from this intellectual effort is a three-part analytical framework. In the first section, the weapons themselves are examined. Here two distinct, but deeply interrelated, questions need to be answered. First, "Have these weapons

10 Lawrence M. Hinman, *Ethics: A Pluralistic Approach to Moral Theory*, 2d. (New York: Thomson Wadsworth, 1998), 165–8.

been as operationally effective and discriminate as has been advertised, thereby reducing human suffering?" and, second, "Have these weapons actually been employed in a manner specifically designed to reduce human suffering to the greatest extent possible?" While this study is not primarily quantitative, weapons effectiveness data and levels of military and civilian casualties should provide valuable insight, particularly as comparatives figures across case studies.

However, questions such as these only address the issue of human suffering during the period of actual combat operations. Unfortunately, the decision to employ PGMs to seek a decisive outcome with lower levels of violence may have a wider impact on human suffering in the longer term. Indeed many urge caution regarding technology's pervasive effect on modern society. Jacques Ellul, in his work *The Technological Society*, is concerned that "technical progress today is no longer conditioned by anything other than its own calculus of efficiency."[11] In an era when technological *means* become the focus, to the relative exclusion of rationally considering *ends*, "certain unforeseen secondary effects ... are much more disastrous than the lack of technique would have been."[12] Approaching from another perspective, the noted naval analyst Norman Friedman, in the article "Is Modern War Too Precise?," wondered if using such precise, partial means, left adversaries "undefeated" psychologically, leading to more resistance and greater human suffering in the long run.[13]

Precisely because of such concerns, the second part of the analytical framework seeks to determine if such second order effects, or unintended consequences, fundamentally alter the conclusions drawn from the first. For example, does the use of such weapons, while reducing initial levels of human casualties, actually result in greater suffering in the long term due to damaged critical infrastructure? Does America's reliance on such high technology solutions force its adversaries to lower tech, asymmetric responses such as suicide bombing which, paradoxically, increase levels of human suffering? Does the use of such weapons result in a more rapid capitulation of the enemy leaving a greater portion of underlying societal and/or political problems unresolved? This second level of analysis is clearly a complex undertaking as it seeks to discern, at some fundamental level, causal relationships between the use of a given technology and wider trends in human suffering observed in societies disrupted by lethal conflict.[14]

The third and final portion of the framework addresses issues at the teleological level. Here the analysis seeks to understand the impact precision has on the national leadership and its decision-making process relative to the use of military

---

11 Jacques Ellul, *The Technological Society*, trans. John Wilkinson (New York: Vintage Books, 1964), 74.

12 Ibid., 105.

13 Norman Friedman, "Is Modern War Too Precise?" *Proceedings* 130.12 (December 2004): 4; Clausewitz stated the results of war are never final (von Clausewitz 1976, 80). Friedman uses the case of Germany following the two World Wars as examples.

14 Econometric analysis, or a "correlates of war" approach, is not the goal of this study.

force. Simply put, how does the use of PGMs, which are a means, influence the consideration of ends? Does the effectiveness of such weapons breed reluctance in American leaders to employ historically more casualty-intensive conventional military forces, thereby opening up opportunities for its adversaries? Perhaps more critically, does having confidence in such weapons, mean senior military and civilian leaders use lethal force more often?

This multi-faceted approach is critical to capturing the full impact of this relatively new technology. Such a holistic approach may highlight areas of concern with current employment philosophies or convince those most concerned with tactical and operational success to consider the longer-range implications of their actions from a human perspective.

Understanding the key definitions and metrics of the study are vital to an understanding of the overall effort. Therefore, for the purposes of this study, the Prussian General Carl von Clausewitz' classic definition of war as "an act of force to compel our enemy to do our will" has been adopted.[15] This deliberately avoids the inherent difficulties of adopting a more legalistic definition of war. The United States has not formally declared war since World War II, although Congress has "authorized" military engagements on six occasions since then.[16] Legalistically, it could be argued that there has never been a use of PGMs during time of "war." Using Clausewitz' definition also permits future research into other types of conflict including non-Western actions.[17]

The term PGMs will be applied only to air-to-surface weapons launched by aircraft, and cruise missiles launched from surface ships or submarines. This is not to imply that other precision technologies do not have an impact, such as any number of high tech anti-armor weapons, or air-to-air missiles. But these latter weapons are primarily tactical in nature, seeking to destroy an individual piece of an enemy's equipment on the battlefield.[18] By design, the study examines those weapons that appear to have the highest possibility, and perception, of delivering operational and strategic results, those at the level of campaigns and above. While, admittedly, such a fine discrimination can be difficult to discern or maintain, it does serve to make the overall project manageable.

---

15 Carl Von Clausewitz, *On War*, ed. and trans. Michael Howard and Peter Paret (Princeton, NJ: Princeton University Press, 1976), 75.

16 See Republican Study Committee Policy Brief titled "Information Regarding Declarations of War and Congressionally Authorized Military Engagements" dated October 23, 2003.

17 Others adopt a more quantitative approach. J. David Singer and Melvin Small, in *The Wages of War, 1816–1965: A Statistical Handbook* (New York: John Wiley and Sons, 1972).

18 Joint Publication 1-02, *Department of Defense Dictionary of Military and Associated Terms*, 12 April 2001 (as amended through 12 July 2007), defines the tactical level of warfare as "the level of war at which battles and engagements are planned and executed ..."

Similarly, the broader category "human suffering" is used intentionally as the object of analysis as opposed to the more commonly used, but more narrowly focused, metric of "casualties." Why is this so? The actual number of people killed and injured during the combat phase of any given war is generally quantifiable to some degree. In fact, one could argue if focused on casualties alone, that modern war is much less deadly than early wars as evidenced by the greatly reduced number of deaths per conflict.[19] But given the scalar differences in both length of time involved and overall level of effort between conflicts such as WWII and the Falklands, direct comparison and meaningful analysis is difficult. Thus, using the more all-encompassing term "human suffering" allows for an analysis of the wider impact of modern war. In addition to deaths and injuries, the term "human suffering" is meant to capture and evaluate the privations a population experiences due to things like the loss of critical infrastructure, the loss of political or economic stability, or the loss of the basic rule of law. At its core, it seeks to measure the loss of a population's ability to flourish in its environment and by design includes both combatants and non-combatants.

Chapter 1 examines the underlying causes and continuum of possible responses to purposeful violence in human societies. This examination focuses on the early synthesis of the Just War Tradition beginning with Augustine and Aquinas and places this tradition in conversation with other efforts, ethical frameworks, international conventions, laws, and religious or cultural traditions, which seek to mitigate human suffering in war.

Chapter 2 examines technology's evolving role in wartime throughout history and introduces a number of key thinkers who raise carefully articulated concerns regarding technology in contemporary society. The French philosopher Jacques Ellul's critique of "*technique*," the drive for ever greater efficiency and the potentially pervasive influences it has on modern society, seems particularly prescient when discussing PGMs.

Chapter 3 examines the underpinnings of a number of ethical traditions, which examine the deeper intellectual underpinnings of the study's just war foundation, seeking to determine if these approaches offers insight on developing metrics for the study. The chapter then outlines the study's underlying methodology and describes the metrics and three-part evaluative framework specifying key indices that measure positive and negative impacts on quality of life as impacted by war.

The next four chapters examine a number of key conflicts in which PGMs played a substantive role. Each of the chapters contains a literature review focused on that particular period, then uses the evaluative framework to analyze the use and impact of PGMs in that conflict. Chapter 4 examines the Vietnam War during the period 1965-1972. While the German World War II-era FX-1400 "Fritz X" and Henschel Hs-293, and early U.S. attempts such as the GB (glide bomb) and VB

19 A press release regarding the UN's first *Human Security Report* released in 2005, reported that in 1950 the average war killed 37,000 people directly but that by 2002 that figure was merely 600.

(vertical bomb) series' weapons mark earlier rudimentary attempts at precision guidance, this period marks the beginning of true operational usage of modern PGMs, albeit as a small part of a much longer war.

Chapter 5 examines the 1990-91 Gulf War, Operations DESERT SHIELD/ STORM. Hailed by many as clear evidence that modern precision warfare was the way of the future, the true story is much more complicated. PGMs constituted a much smaller percentage of munitions expended than is popularly realized. Similarly, the flat, open environment of Iraq was an ideal proving ground for such weapons. Here first and second order effects are evident.

Chapter 6 focuses on the two interventions in the Balkans, Bosnia in 1995, and Kosovo in 1999. In particular, the study will examine the offensive actions of Operation DELIBERATE FORCE (Bosnia), and Operation ALLIED FORCE (Kosovo). What is significant about these operations is that U.S. and NATO leaders relied almost exclusively on air power, in its various forms, in an attempt to influence the political actions of its adversaries. The true landscape is, of course, much more complicated and many factors, including pressure on critical support structures, and the lessening of external political support for the regime, contributed to the eventual capitulation. But by relying heavily on PGMs, the direct and second order effects emerge in clearer focus during these actions due to a lack of the intervening variable of ground forces. These case studies also offer keen insight at the teleological level.

Chapter 7 approaches the overarching question from a different perspective. During the more than three-decade period under study, the U.S. government has launched combat operations utilizing PGMs, designed to influence another government's actions, on a number of occasions. These coercive operations include operations into Libya (twice), Iraq, the Persian Gulf, Sudan, and Afghanistan. These shorter operations may provide valuable insight into the more teleological aspects of the analysis. The final chapter contains the summary, including reflections on more contemporary operations, and the study's conclusions.

That PGMs, clearly an advanced technological weapon, grant the possessor a highly efficient and effective capability to eliminate fellow human beings with limited risk to his/her own forces is a demonstrated fact. That we "can," however, does not mean we "should." How we deal with this reality, particularly from an ethical perspective, is where the humanities enter the equation. To-date the emphasis appears to be on reducing human suffering by the use of PGM's highly discriminate targeting capability and extremely careful planning to reduce civilian casualties and collateral damage.[20]

20 Note: The views expressed in this book are solely those of the author and do not represent or reflect the position or endorsement of any governmental agency or department, military or otherwise.

Chapter 1

Purposeful Violence: Underlying Causes and the Continuum of Human Responses

> Never, never, never believe any war will be smooth and easy, or that anyone who embarks on that strange journey can measure the tides and hurricanes he will encounter.
>
> Winston Churchill, A Roving Commission

Evidence of intentional violence by human beings against one another is found throughout the historical and archaeological record. In "hunter-gatherer" times, successful competition for scarce resources meant the continued viability of a small clan or group, at the expense of other clans or groups in the local area. While speaking of primitive man, Bertrand Russell noted, "early man must have lived in small groups not very much larger than families ... Within each group there seems to have been a considerable amount of cooperation, but towards all other groups of the same species there was hostility whenever contact occurred."[1] Such behavior is still evident today, exhibited by a number of "isolated," or "uncontacted," indigenous tribes, where simple encroachment on their claimed territory can, and does, result in a swift, violent, and gruesome death for the outsider.[2] The Greek historian Thucydides also offers an anthropological analysis of the transitory nature of early clan life. In the opening portion of his history of the Peloponnesian War known as "the archaeology," he writes "it is evident that the country now called Hellas had in ancient times no settled population; on the contrary, migrations were of frequent occurrence, the several tribes readily abandoning their homes under the pressure of superior numbers."[3]

As societal structures changed, so too did the reasons for fighting between rival groups. With clans and tribes settling in agrarian villages during the Neolithic age, and city-states during the Bronze, violence evolved into a tool for gaining comparative advantages for your city-state, often when other means had failed.[4]

---

1 Bertrand Russell, *Authority and the Individual: The Reith Lectures for 1948–9* (London: Unwin Brothers Ltd., 1949), 13.

2 See Paul Raffaele's "Out of Time," *Smithsonian Magazine* April 2005: 62–70.

3 Thucydides, *The Landmark Thucydides: A Comprehensive Guide to the Peloponnesian War*, edited by Robert B. Strassler (New York: The Free Press, 1996), 3.

4 Grahame Clark, "The Beginning of Farming: Jericho and Jarmo," *The World of the Past*, ed. Jacquetta Hawkes (New York: Alfred A. Knopf, 1963), 249–53.; John B. Harrison and Richard E. Sullivan, *A Short History of Western Civilization*, 4th ed. (New York: Alfred A. Knopf, 1975), 6–8.

Martial contests, originally designed to exterminate the opponent, became battles of conquest where "the vanquished, instead of being put to death, were made slaves and compelled to labour for their conquerors."[5] In this context, violence between societies became a purpose-driven activity, and ever more recognizable as what we today refer to as "war."

Thucydides again provides insight into this human desire to improve its relative position *vis-à-vis* others through conflict, stating:

> no one can quarrel with a people for making, in matters of tremendous risk, the best provisions that it can for its interest ... it was not a very remarkable action, or contrary to the common practice of mankind, if we did accept an empire that was offered to us, and refused to give it up under the pressure of three of the strongest motives, fear, honor, and interest. And it was not we who set the example, for it has always been the law that the weaker should be subject to the stronger.[6]

From an Eastern perspective, the ancient Chinese general and military theorist Sun Tzu echoed a similar sentiment stating, in the first sentence of his book *The Art of War*, "war is a matter of vital importance to the State; the province of life or death; the road to survival or ruin."[7] Against the backdrop of the "Warring States Period," he employed a broader definition than most theorists, believing that states were always at war—in a constant state of competition. Interestingly, the pinnacle of skill, he believed, was to win this competition "without fighting."[8] Later, Carl von Clausewitz, in his seminal work *On War*, adopted a narrower approach by declaring that war was "an act of force to compel our enemy to do our will."[9] Clearly, significantly different historical and cultural settings conditioned these author's writings. However, there is also much commonality here, what Michael Handel perceived as a "universal logic" to strategy and war.[10]

While the propensity for violence is deeply engrained in the human psyche, what we are examining here is that which is manifested at the national level. It is worth repeating that nations exhibit behavior in diplomacy and statecraft that individuals find morally repulsive at the personal level, the subject of Reinhold Neibuhr's work *Moral Man and Immoral Society*.[11] Optimists might still hope

5 Russell, *Authority and the Individual*, 16.

6 Thucydides, *Landmark Thucydides*, 43.

7 Sun Tzu, *The Art of War*, translated by Samuel B. Griffith (Oxford: Oxford University Press, 1963), 63.

8 Ibid., 77.

9 Clausewitz, *On War*, 75.

10 Michael I. Handel, *Masters of War: Classical Strategic Thought*, 3rd ed. (London: Frank Cass, 2001), xvii.

11 Reinhold Niebuhr, *Moral Man and Immoral Society: A Study in Ethics and Politics*, (New York: Charles Scribner's Sons, 1932).

that one day humankind will reach the point where "nation will not lift up sword against nation, and never again will they learn war."[12] That said, whether the concept of war is viewed broadly or more narrowly, as evidenced by the recent and ongoing military operations in Iraq, Afghanistan, and elsewhere, we live in a world where armed conflict is generally considered a legitimate, if regrettable, tool of statecraft. It is a harsh reality that, for reasons both good and bad, nations sometimes engage each other in lethal combat to right perceived wrongs or gain a comparative edge. As Art and Waltz observe, "conflict marks all human affairs."[13] How then should humankind respond to this moral quandary, and what role does technology play in that response?

Humankind's Response to Purposeful Violence: A Continuum

At one end of the response spectrum is the ideology of in-principle pacifism, the belief that violence or the use of force, and by extension war, is not justified in any situation. Such pacifists include the Religious Society of Friends (Quakers), the Amish, and the Mennonites. Indeed, a Christian scriptural basis for this position is in Jesus' Sermon on the Mount where He stated: "but I say to you, do not resist an evil person; but whoever slaps you on your right cheek, turn the other to him also."[14] Pacifism does not equate, however, to a lack of response but seeks resistance by other means, a position adopted by persons of such varied cultural backgrounds as Mahatma Gandhi, Thomas Merton, and Thich Nhat Hanh.[15]

While this voice must not be lost, serving to moderate those on the more aggressive end of the continuum, it fails to address pragmatically what should be done when your nation is under lethal threat. Indeed, Gandhi received much criticism for comments he made concerning the plight of the Jews during World War II, when he argued that, although "the German persecution of the Jews seems to have no parallel in history ... I do not believe in any war."[16] While not meant to imply a consequentialist foundation to the belief, there is a conviction here that the costs of war far outweigh any benefits, even if you win. As Erasmus cautioned, "you will lose much more than you gain."[17] Interestingly, not all pacifists believe

12 Isa. 2:4 NASB

13 Robert J. Art and Kenneth N. Waltz, "Technology, Strategy, and the Uses of Force," *The Use of Force: International Politics and Foreign Policy*, ed. Robert J. Art and Kenneth N. Waltz (New York: University Press of America, 1983), 1.

14 Mt. 5:39 NASB

15 See Robert King's *Thomas Merton and Thich Nhat Hanh: Engaged Spirituality in an Age of Globalization.* (New York: Continuum, 2001).

16 David Lewis Schaefer, "What Did Gandhi Do? One Sided Pacifist," *National Review Online* 23 April 2003, 19 September 2006 http://www.nationalreview.com/comment/comment-schaefer042803.asp.

17 Mark R. Amstutz, *International Ethics: Concepts, Theories, and Cases in Global Politics* (New York: Rowman & Littlefield, 1999), 94–6.

that force and violence should never be used. Bertrand Russell, for example, argued that in the case of defeating Hitler, war was not the worst of all possible evils. Similarly, there were Buddhists in Sri Lanka that argued for a military solution to the ongoing ethnic problems there.[18]

At the other end of the conceptual response spectrum is what Mark Amstutz describes as the ideology of "amoral realism." He explains:

> Whereas pacifism prohibits war and violence, amoral realism assumes not only that war is a legitimate instrument of policy but also that morality does not constrain war. Indeed, assuming the legitimacy of war's ends, amoral realism believes that the only norm governing the conduct of war is military victory. Moreover, because morality is absent in the international community, global order is achieved not by the pursuit of "moral" foreign policies but rather as a direct by-product of the balance of power among states.[19]

Those grounded in international relations theory will recognize the "realist" school of thought, generally seen as the "dominant theoretical tradition of the Cold War."[20] This theory, first evidenced by Thucydides when describing Athens' dealings with the small island of Melos, and advanced by Machiavelli in the sixteenth century, is perhaps best captured by one of its most notable modern adherents, Hans Morgenthau, who argued:

> The actions of states are determined not by moral principles and legal commitments but by considerations of interest and power. Moral principles and legal commitments may be invoked to justify a policy arrived at on other grounds ... but they do not determine the choice among different courses of action.[21]

As stated here, the realist position could be seen as reflecting a sort of societal or national egoism, the ethical framework advanced by the British philosopher Thomas Hobbes that rests upon the proposition that humans act in their own self-interest without regard to others. Taken to a logical extreme, Lawrence Hinman's

18 Harris, "Buddhism," 94.

19 Amstutz, *International Ethics*, 96–7.

20 Stephen M. Walt, "International Relations: One World, Many Theories," *Conflict and Cooperation: Evolving Theories of International Relations*, ed. Mark A. Genest (Belmont, CA: Thomson Wadsworth, 2004), 30.

21 For discussion on Thucydides and Machiavelli see Thomas L. Pangle and Peter J. Ahrensdorf, *Justice Among Nations: On the Moral Basis of Power and Peace* (Lawrence: University Press of Kansas, 1999) 13–18, 125–37; See also, Walzer, *Just and Unjust War*, 5–12.; Morgenthau quoted in Ward Thomas, *The Ethics of Destruction: Norms and Force in International Relations* (Ithaca, NY: Cornell University Press, 2001), 5.

definition of egoism could be reworked such that "egoism maintains that [states] *should* always act in their own self-interest" without regard to other states.[22]

James Turner Johnson believes "realism was well equipped to deal with the stable bipolar reality of the cold war era."[23] But he also argues that the international order has changed greatly since the end of the Cold War and such a change demands a new approach to addressing the use of power, one based on moral wisdom.[24] Adding to this viewpoint, Amstutz and others argue that, while power and necessity are certainly key factors in any nation's foreign policy decisions, in opposition to the realist position, morality and choice do and should play a role in this arena.[25] Any number of events in the recent past seem to confirm this assertion as active external intervention in the former Yugoslavia, Haiti, Somalia (particularly in the early famine relief operations), and perhaps more recently Libya were, at least in part, driven by international moral condemnation of human suffering.[26]

Located between these polar "never" and "always" extremes, is an intermediary position Amstutz calls "principled realism." This approach rests on the assumption that:

> all interstate relations, including war, are subject to widely acknowledged moral standards. Because power and force are fundamental realities of the international community, the quest for peace and justice among states can only be imperfectly realized, and then only when power and force are incorporated into the calculus of decision making. Like pacifism, principled realism affirms international moral constraints on war, but whereas pacifism proscribes all war and violence, principled realism assumes force can be an instrument of justice.[27]

Here, perhaps, is a pragmatic middle ground. Fundamental to this definition is the realization that the use of violence to solve problems, although seen as suboptimal, is not necessarily the "greatest evil." Significantly, "just as an individual has the right to resist injustice, so too a state has the right to use force, individually or collectively, to repulse aggression and to resist and punish evil."[28] But where are the foundations of Johnson's "moral wisdom" or Amstutz' "moral standards" to be found?

---

22 Hinman, *Pluralistic Approach*, 121.

23 James T. Johnson, *Morality and Contemporary Warfare* (New Haven: Yale University Press, 1999), 20.

24 Ibid., 21–2.

25 Amstutz, *International Ethics*, xii. See also Pangle and Ahrensdorf, *Justice Among Nations* and Thomas, *Ethics of Destruction.*

26 Ibid., xii–xiii.

27 Ibid., 99.

28 Ibid., 100.

Navigating the Middle Ground

Historical evidence indicates that humankind has actively sought ways to justify and limit the resort to war, and to limit conduct in war, for thousands of years. Much of the reasoning behind such limitations sought to reduce or mitigate human suffering. Such measures have taken the form of ethical frameworks, religious or cultural traditions, and international laws and conventions. The most famous and influential of Western ethical traditions, and the one Johnson believes in most deeply, is *justum bellum*, or the Just War Tradition. Today, just war-influenced thought can be found embedded throughout existing international laws and conventions.

However, Western thinking alone does not begin to capture the totality of human reflection on this important issue. While perhaps couched in different language, and sometimes difficult to decipher through a Western lens, cultures and religions around the world have also sought to determine when it is permissible to resort to the use of violence and what conduct is permissible in war.[29]

Ethical Frameworks: The Just War Tradition—A Western Approach

Building upon earlier Greek and Roman classic writings and the foundational works of Christian theologians such as Saint Augustine (A.D. 354–430) and Saint Thomas Aquinas (A.D. ca. 1227–1274), Christian and secular philosophers have sought to develop a framework for warfare, an effort that for much of its history has principally evolved along two critical branches. The first, *jus ad bellum*, seeks to determine when it is morally correct to go to war. The second, *jus in bello*, seeks to determine what actions are morally permissible in war. A third branch, *jus post bellum*, newer and, to this point, less well-developed, deals with correct actions after the "just war" is over. The Just War Tradition can be conceived of in a number of ways.

The American Catholic Bishops see it as "a moral theory with roots in both natural law and in Christian theology, but developed and given definitive form by Catholic moral teaching."[30] The Protestant theologian Paul Ramsey, a major voice in the just war debate from the 1950s until his death in 1988, saw the Just War Tradition as arising "from the common moral wisdom of humankind."[31] Similarly, Michael Walzer, from a secular perspective, understands just war as "a set of ideas generated out of common human experience of the reality of war and reflection on that experience."[32] However one comprehends just war, it is clear that the tradition has weathered momentous changes, both in terms of the structure of society and

29 *Just War in Comparative Perspective*, ed. Paul Robinson (Burlington, VT: Ashgate, 2003) examines how a number of nation/cultures/religions treat "just war."

30 Johnson, *Contemporary Warfare*, 22.

31 Ibid., 23.

32 Ibid.

in the lethality of weapons. Yet, despite these changes, its underpinnings have remained remarkably consistent for 1,600 years.[33]

Jus Ad Bellum

Although Saint Ambrose (A.D. 340–397) did earlier foundational work on the virtue of justice, particularly related to making battlefield decisions, most acknowledge Augustine as the progenitor of just war theory.[34] Augustine's fundamental beliefs draw heavily upon the classic writings of Plato and Marcus Tullius Cicero. Plato, for example, provides an early philosophical basis for right authority in the *Laws*:

> Any person making peace or war with any parties independently of the Commonwealth shall likewise incur the pain of death. If a section of the state makes peace or war with any on its own account, the generals shall bring the authors of the measure before a court, and the penalty for conviction shall be death.[35]

Cicero, too, provides philosophical precursors in *De Re Republica* 3, XXIII, stating:

> A war is never undertaken by the ideal state, except in defense of its honor or safety …
>
> Those wars which are unjust are undertaken without provocation. For only a war waged for revenge or defense can actually be just. …
>
> No war is considered just unless it has been proclaimed and declared, and unless reparation has first been demanded.[36]

In *De Officiis*, Cicero also addresses the issue of war as a course of last resort, to be used only when discussion was unsuccessful.[37]

33 Ibid., 1.

34 Paul Pasquale Christopher, *The Ethics of War and Peace: An Introduction to Legal and Moral Issues*, 3rd ed. (Upper Saddle River, NJ: Pearson Prentice Hall, 2004), 23–4, 29; Brother John Raymond, *The Just War Theory*, 26 September 2006 http://www.monksofadoration.org/justwar.html.

35 Excerpted from Plato, *Laws*, Bk. XII 955b9-c6, in *The Collected Dialogues of Plato*, edited by Edith Hamilton and Huntington Cairns (Princeton, N.J.: Princeton University Press, 1961) quoted in Christopher, *Ethics of War and Peace*, 10.

36 Excerpted from Marcus Tullius Cicero, *De Re Republica* 3, XXIII, translated by Clinton Walker Keyes (New York: G.P. Putnam's Sons, 1928), 211–13, quoted in Christopher, *Ethics of War and Peace*, 12.

37 Ibid.

If he drew so heavily on earlier thinkers, and if, as Paul Christopher suggests, his contemplations on just war are widely interspersed throughout his writings making it difficult to find "a single coherent theory," then just what is Augustine's lasting contribution?[38] The answer is that his synthesis of earlier Greek and Roman philosophers, and the evolving tenets of early Christianity, forms the foundation upon which later religious and secular writers build. Indeed, perhaps his greatest contribution to the tradition is rooting it in the understanding, expressed in his A.D. 418 *Letter to Count Boniface*, that "war is waged in order that peace may be obtained."[39]

One of the most important events in the development of the tradition occurred in 1148 when a Benedictine monk, Gratian, published the *Concordia discordantium canonum*, an exhaustive compilation of canon law. More commonly known as the *Decretum*, this work was critical because it centralized the debate for future philosophers—such as Saint Thomas Aquinas.[40] While Aquinas' basic work on *jus ad bellum* considerations was not new, "he was the first to place all three of them together as independently necessary and jointly sufficient."[41] In *Summa Theologica* II, Q. 40, Aquinas wrote:

> In order for a war to be just, three things are necessary. First, the authority of the sovereign by whose command the war is to be waged. ... Secondly, a just cause is required, namely that those who are attacked, should be attacked because they deserve it on account of some fault. ... Thirdly, it is necessary that the belligerents should have a rightful intention, so that they intend the advancement of good, or the avoidance of evil.[42]

Through the Middle Ages, the language and locus of the Just War Tradition were firmly rooted in the Catholic faith. As Europe exited the sixteenth century, society was confronted by the rise of the nation-state, an increase in colonialism, the multifaceted Protestant Reformation, and the loss of relative power for the Catholic Church. During this turbulent period, the next major evolution in the Just War Tradition occurred, the transformation of the conversation from one solely of Catholic theologians into one that included thinkers from the secular realm, a movement led by the Dutch jurist Huig de Groot, better known as Hugo Grotius.

Influenced by the ongoing horrors of the Thirty Years War, a war in which "the only outcome acceptable to the opponents was either the conversion or the

38 Ibid., 30.

39 Philip Schaff, *The Confessions and Letters of St. Augustin, with a Sketch of His Life and Work* (New York: Christian Literature Publishing Co., 1886), 554.

40 Christopher, *Ethics of War and Peace*, 48.

41 Ibid., 50.

42 St. Thomas Aquinas, *Summa Theologica*, Christian Classics Ethereal Library, no date, Calvin College, 27 September 2006 http://www.ccel.org/a/aquinas/summa/SS/SS040.html#SSQ40OUTP1.

annihilation of the opposing side," Grotius published *De Jure Belli ac Pacis* or *On the Law of War and Peace* in 1625.[43] He explained:

> I have had many and weighty reasons for undertaking to write upon this subject. Throughout the Christian world I observed a lack of restraint in relation to war, such as even barbarous races should be ashamed of; I observed that men rush to arms for slight causes, or no cause at all, and that when arms have once been taken up there is no longer any respect for law, divine or human; it is as if, in accordance with a general decree, frenzy had openly let loose for the committing of all crimes.[44]

Grotius sought to decouple the discussion from what he believed was a corrupt church and replace it with "an external, objective, secular authority that the competing political interests (i.e. nation-states) would accept—a corpus of international laws."[45] These efforts, Paul Christopher believes, "represent the culmination of a thousand years of gradually reducing moral principles to objective criteria."[46] Not surprisingly, this shift in legal emphasis roughly corresponds with the wider decline of "the automatic legitimacy of sacral monarchy" that ultimately results in the rise of the concept of nationalism.[47]

Jus ad bellum criteria Based on both religious and secular foundations, today six criteria form the core of most modern *jus ad bellum* interpretations, although some authors alter them as their individual views warrant. James Turner Johnson, for example, adds "the aim of peace" as a seventh criterion. While clearly an early concern articulated by Augustine, this criterion seems sufficiently embedded in the definition of the principle of "right intentions."[48] The first *jus ad bellum* principle is that of just cause, which states "war may be waged only to meet a real and grave danger. Such danger threatens innocent human life or basic human rights."[49] Next is the principle of right intention. Since war may only be fought to meet a real and grave danger, "a war may be fought only long enough to realize just goals; a nation must avoid unnecessary brutality or engaging in practices which would threaten the establishment of a just, lasting peace."[50]

---

43 Christopher, *Ethics of War and Peace*, 67.

44 Ibid. Original: Hugo Grotius, *The Law of War and Peace*, translated by Francis W. Kelsey (Indianapolis: Bobbs-Merrill Co., 1962), 20.

45 Ibid.

46 Ibid., 81.

47 Anderson, *Imagined Communities*, 21.

48 Johnson, *Contemporary Warfare*, 29.

49 John D. Jones and Marc F. Griesbach, eds., *Just War Theory in the Nuclear Age* (New York: University Press of America, 1985), viii.; Also Johnson, *Contemporary Warfare*, 28–9.

50 Ibid.

A third is the principle of *legitimate authority*. "War can be waged only to serve public interest or the common good and not private gain. Therefore, only those public officials who have the right or duty to declare war may so declare it."[51] The principle of *last resort* requires that "a war may be waged only if all peaceful means to avoid war have been exhausted."[52] Additionally, except for some wars of self-defense, a just belligerent must have a *reasonable chance of success* in waging the war. Finally, the principle of *proportionality* demands that "the destruction and evils brought on by war must be proportionate to the just causes for which the war is fought and the goods to be achieved by waging the war."[53]

Jus In Bello

Aquinas was also the first to properly explain the concept of "double effect," which offers a justification for killing civilians in war, so long as their deaths are not directly intended. Aquinas used this to justify self-defense, a clear break from Augustine, but its effects have had a far ranging impact on a variety of aspects of warfare.[54] Since much of this study's later framework will rest upon the issue of intended versus unintended consequences, it is important that the underlying conditions germane to this concept be clearly articulated. Under double effect, the killing of noncombatants would be permitted provided four conditions were met. As Walzer outlines, these are:

1. The act is good in itself or at least indifferent, which means, for our purposes, that it is a legitimate act of war.
2. The direct effect is morally acceptable—the destruction of military supplies, for example, or the killing of enemy soldiers.
3. The intention of the actor is good, that is, he aims only at the acceptable effect; the evil effect is not one of his ends, nor is it a means to his ends.
4. The good effect is sufficiently good to compensate for allowing the evil effect; it must be justifiable under [the] proportionality rule.[55]

Despite this early excursion into *jus in bello* concerns, until the sixteenth century, most intellectual effort was focused on *jus ad bellum* considerations. Then, Francisco de Vitoria, while also offering new insights into proportionality and the need to use force only as a last resort, expanded thinking in a new direction.

51 Ibid.

52 Ibid., viii–ix.

53 Ibid., ix.

54 Christopher, *Ethics of War and Peace*, 52–3.

55 Michael Walzer, *Just and Unjust Wars: A Moral Argument with Historical Illustrations*, 3rd ed. (New York: Basic Books, 2000), 153.

Affected by the treatment of indigenous American natives by the Spanish Conquistadors, he sought to determine whether the ongoing massacres by the Spanish in the New World were right or wrong. In doing so Vitoria significantly expanded just war thinking in the *jus in bello* realm. Today, two principles form the core of this branch of the Just War Tradition.

Jus in bello criteria The first *jus in bello* principle closely resembles the final *jus ad bellum* principle, namely that proportionate means be used. In this case however, "the calculus of 'proportion' hinges more narrowly on the legitimate military goals to be achieved, the forces arrayed on the enemy side, and the least destructive ways to defeat those forces or render them ineffective so as to achieve those legitimate ends."[56] Additionally, the principle of discrimination "requires that noncombatants should not be directly, intentionally targeted, even in the course of using force that is otherwise proportionate."[57]

Jus Post Bellum

In *The Three Theological Virtues*, the Spanish theologian Francisco Suárez wrote, "three periods must be distinguished with respect to every war: its inception; its prosecution, before victory is gained; and the period after victory."[58] To date almost all development in the Just War Tradition has addressed the first two periods. More recently, discussions regarding actions during the third, post-war, period have emerged—a period in which *jus post bellum* concerns become critical. Perhaps, as the world experiences any number of turbulent post-conflict periods, attention needs to be focused on this emerging branch of the theory.

Davida Kellogg, for example, believes that war crimes trials are a vitally important part of the post-war period. In making her case, she argues:

> If Just War is undertaken to right wrongs done by a group or groups of people to another ... then stopping short of trying and punishing those most responsible for war crimes and crimes against humanity which either led to war or were committed in its prosecution may be likened to declaring 'checkmate' and then declining to take your opponent's king. It makes no strategic sense, since the purpose for which war was undertaken is never achieved. It makes no legal sense, since the criminal activities the war was undertaken to rectify or curtail are allowed to continue unchecked. What is worse, it makes no moral sense, since justice is not done for the victims of atrocities in such an outcome to war.[59]

---

56 James T. Johnson and George Weigel, *Just War and the Gulf War* (Washington, D.C.: Ethics and Public Policy Center, 1991), 31.

57 Ibid.

58 Quoted in Christopher, *Ethics of War and Peace*, 58; Davida E. Kellogg, "Jus Post Bellum: The Importance of War Crimes Trials," *Parameters* XXIII.3 (Autumn 2002): 87.

59 Kellogg, "War Crimes Trials," 88.

Rear Admiral (ret.) Louis V. Iasiello, former Chief of the U.S. Navy Chaplain Corps, argues that *post bellum* activities need to be guided by both legal *and* moral precepts. While the Fourth Geneva Protocol of 1949 and the Hague Convention IV of 1907 address the legal aspects of the post-war phase, he finds similar moral precepts lacking. He offers a proposed framework of seven parameters: a healing mind-set, just restoration, safeguards for the innocent, respect for the environment, *post bellum* justice, the transition of warriors, and the study of the lessons of war. Adhering to these seven parameters, he believes, "may ultimately save lives and enhance chances to secure a just and lasting peace for all."[60]

One of the most clear and persuasive advocates regarding the need for the greater development of practical principles for Suárez' third period is Brian Orend, who credits Immanuel Kant with being "the first great thinker to stress, in a meaningful and substantive fashion, the topic of *jus post bellum*."[61] In *The Doctrine of Right*, Kant wrote:

> International right [or justice] is thus concerned partly with the right to make war, partly with the right of war itself, and partly with the questions of right after war, i.e. with the rights of states to compel each other to abandon their war-like condition and to create a constitution which will establish an enduring peace.[62]

Comparing the quantity, and substance, of *jus post bellum*-related language in international laws and conventions, such as the Hague and Geneva Conventions, and the United Nations (UN) Charter, with that of *jus ad bellum*- and *jus in bello*-related material, Orend concludes, "there is a gross imbalance in the international laws of armed conflict. And it is war termination that gets the short end of the stick."[63] In an effort to advance the conversation in post-conflict just war thinking, Orend proposes five *jus post bellum* principles, carefully written to reflect current just war language, and which broadly cover the concerns of Kellogg and Iasiello.

Proposed jus post bellum principles The first *jus post bellum* principles is "just cause for termination." Under this principle, a state has just cause for termination if, simply stated, the wrongs that justified the just war in the first have been vindicated or resolved. He continues:

> Not only have most, if not all, unjust gains from aggression been eliminated and the objects of Victim's rights been reasonably restored, but Aggressor is now

60 Rear Admiral Louis V. Iasiello, "Jus Post Bellum: The Moral Responsibilities of Victors in War," *Naval War College Review* 57.3/4 (Summer/Autumn 2004): 51.

61 Brian Orend, "Jus Post Bellum," *Journal of Social Philosophy* 31.1 (Spring 2000): 118.

62 Quoted in Orend. Ibid.

63 Brian Orend, *War and International Justice: A Kantian Perspective* (Waterloo, Ont.: Wilfrid Laurier University Press, 2000), 220.

> willing to accept terms of surrender which include not only the cessation of hostilities and its renouncing the gains of its aggression but also its submission to reasonable principles of punishment, including compensation, *jus ad bellum* and *jus in bello* war crimes trials, and perhaps rehabilitation.[64]

The second principle is that of "right intention." The victorious state must "intend to carry out the process of war termination only in terms of those principles contained in the other jus post bellum rules."[65] Under this principle, revenge should not be a motivating factor and, if there are to be war crimes trials, the victor must equally apply to both sides the rules of law in terms of investigation and prosecution.[66]

The next principle is "public declaration, legitimate authority and domestic rights protection." Here, "the terms of the peace must be publicly proclaimed by a legitimate authority ... and domestic rights must be fulfilled just as readily as external rights."[67] The fourth principle is "discrimination." Within the aggressor state, the victor must differentiate between those who lead the war, civilian and military leaders, those who actively participated in the war, the soldiers, and the civilian population. Critically, "undue and unfair hardship is not to be brought upon the civilian population in particular: punitive measures are to be focused upon those elites most responsible for the aggression."[68] The fifth and final principle Orend proposes is that of "proportionality." Echoing sentiments found in both *jus ad bellum* and *jus in bello* criteria, "any terms of peace must be proportional. ... Absolutist crusades against, and/or draconian punishments for, aggression are especially to be avoided. The people of the defeated [state] never forfeit their human rights, and so are entitled not to be 'blotted out' from the community of nations."[69]

The Just War Debate in the Nuclear and Precision Eras

In *Just and Unjust Wars: A Moral Argument with Historical Illustrations*, Michael Walzer develops an argument for what he calls a "supreme emergency," a danger so great that it "may well require exactly those measures that the war convention bars."[70] Walzer outlines two criteria to be used when making such a determination, imminence and the nature of the danger, clearly stating that neither is sufficient by itself. He believes the early days of World War II, following the fall of France, satisfied these conditions and therefore justified, at least for a time, the intentional

64 Ibid., 232.
65 Ibid.
66 Ibid.
67 Ibid "Legitimate authority" defined on p. 112.
68 Ibid.
69 Ibid., 232–3.
70 Walzer, *Just and Unjust Wars*, 251. Winston Churchill, first used the term in 1939.

bombing of civilians in German cities by the British.[71] However, not all agree with Walzer's interpretation. Martin Cook and Brian Orend argue instead that there should be no exceptions to *jus in bello* rules. Violations by political leaders are violations, but may, in very rare cases, be forgivable after the fact.[72]

Whether one accepts the Walzer's interpretation or not, it must be acknowledged that innocent civilians, normally protected from direct attack, were targeted during the blitz in London, the fire-bombings of Tokyo and Dresden, and the atomic attacks in Japan, most of which did not satisfy the supreme emergency criteria in any case.[73] The aftereffects of these actions were significant. David Hollenbach is convinced that "once noncombatant immunity is set aside in a situation of 'supreme emergency,' a social practice is legitimated that will be very difficult to delegitimate in the future" and that "the decision to resort to strategic bombardment during World War II provided a precedent for the development of strategies of counterpopulation nuclear warfare."[74]

In reaction to the development of such strategies in the Cold War, the writings of John Courtney Murray, Paul Ramsey, and James Turner Johnson rekindled a hitherto dormant just war debate.[75] The ensuing arguments were robust, passionate, and at times complex in nature. Here the debate was more about means than ends, focusing primarily on the overwhelming destructive power of the weapons themselves. Some opponents of nuclear weapons argued from a *jus ad bellum* principle of "reasonable chance of success" perspective. They believed that "the use of nuclear weapons ... would escalate into an unlimited ... war ... [and] if nuclear weapons cannot be limited in their use ... there can be no ... reasonable hope for success."[76] Others felt that "any nuclear war, no matter how limited in intent, would result in a world in which the political goals of its initiators would be irrelevant and hateful to its pitiful survivors."[77]

Greater concerns, however, were voiced in relation to *jus in bello* considerations. "Given the destructiveness, hazards of radioactivity, and dangers of escalation inherent in recourse to nuclear weapons, a general presumption exists against their proportionality if nonnuclear means of commensurate effectiveness are available."[78]

---

71 Ibid., 252–5.

72 Martin L. Cook, "Michael Walzer's Concept of 'Supreme Emergency,'" *Journal of Military Ethics* 6.2 (2007): 138–51.

73 Arguments in favor of such bombing have been raised.

74 David Hollenbach, "Ethics in Distress: Can There Be Just Wars in the Nuclear Age?" *The Nuclear Dilemma and the Just War Tradition*, ed. William V. O'Brien and John Langan (Lexington, MA: Lexington Books, 1986), 17.

75 J. Bryan Hehir, "Just War Theory in a Post-Cold War World," *Journal of Religious Ethics* 20.2 (Fall 1992): 239–40.

76 Keith B. Payne and Payne, Karl I., *A Just Defense: The Use of Force, Nuclear Weapons, and Our Conscience* (Portland, OR: Multnomah, 1987), 49.

77 Jones and Griesbach, "Just War Theory," 85.

78 William V. O'Brien, *The Conduct of Just and Limited War* (New York: Praeger, 1983), 132.

Despite his caveat regarding conventional means, O'Brien captures the thrust of the proportionality argument. Here the magnitude of the instrument's destructiveness, coupled with the unavoidable collateral damage that would inevitably occur, renders the use of nuclear weapons, for many, fundamentally immoral.[79]

Other critics used the principle of discrimination as their touchstone. Nuclear pacifists led this effort with the argument that "nuclear weapons are ... indiscriminate in their destruction, failing to distinguish between the innocent and the evil."[80] Others pointed to the inaccuracy of delivery vehicles. "[Since] pinpoint targetting [*sic*] against military objectives, particularly silos, was not possible ... the destruction of Moscow could be assured ... no matter into which Moscow suburb the missiles fell."[81] Such concerns were difficult to assuage as both counter-force and counter-value strategies held the promise of thousands, if not millions, of civilian deaths in the event of a nuclear exchange.[82]

During this period, most argued that conventional war would tend, almost inevitably, towards nuclear war, and thus by extension *jus in bello* considerations would become moot. The employment of PGMs, offering the possibility of attaining objectives more quickly and cleanly, may permit conventional war to remain conventional.[83] Indeed, the introduction of PGMs into the American arsenal, and their demonstrated effectiveness particularly since 1991, has alleviated many of the concerns of those critics arguing from *jus in bello* considerations.

James Turner Johnson offers an authoritative voice when speaking about the need to make an intellectual transition, to alter conventional thinking about contemporary warfare, arguing:

> PGMs give the American military ... the ability to fight in the way moralists have long been saying they should fight: in a way that avoids harm to noncombatants and minimizes overall destruction to the society which, after the war is over, must be brought back to a state of peace. Anyone who has followed the American moral debate over military force for the last forty or so years knows that a recurring theme among opponents of military force has been that, in the contemporary age, war is inherently indiscriminate and disproportionately destructive and that any use of military force is accordingly questionable. ...

---

79 Jones and Griesbach, *Just War Theory*, 81; Payne and Payne, *Just Defense*, 49–50; Walzer, *Just and Unjust Wars*, 277.

80 Payne and Payne, *Just Defense*, 49.

81 Jones and Griesbach, *Just War Theory*, 81.

82 *Counter-force* strategies target military capability, *counter-value* strategies target civilians and cities.

83 In the early 1980s, a shift away from the former belief began to emerge among American policy makers, a trend which Barry Posen called "dangerous and unwarranted." See Barry R. Posen, "Inadvertent Nuclear War: Escalation and NATO's Northern Flank," *Strategy and Nuclear Deterrence: An International Security Reader*, ed. Steven E. Miller (Princeton: Princeton University Press, 1984), 85.

> My point is simple and to the contrary: Contemporary war is not inherently indiscriminate or grossly destructive. It is not by necessity nuclear war or even warfare characterized by World War II-type strategic bombing. The use of military force does not, because of some inherent law, escalate to such levels or "spin out of control," as one critic of the Gulf War put it. Rather, *PGMs provide a vastly increased operational capability to fight according to the moral requirement of the just war idea.*[84]

Network news footage showing pinpoint accuracy, demonstrated weapons effectiveness rates approaching 90 percent, and careful planning by military staff members clearly address the principles of proportionality and discrimination. Writing about Afghanistan, Peter Steinfels notes, "it already seems indisputable that the United States military not only rejected direct attacks on civilians but also strove mightily to avoid what is antiseptically termed 'collateral damage' ... [and that PGMs] have supplied new means for discriminating between military and civilian targets."[85]

This does not imply that all are happy with this new manner of warfare for "as precision increases, so do expectations, constantly 'raising the bar' of acceptable performance. As a result, tolerance for inaccuracy or even human error diminishes."[86] The attack on the *al Firdos* command and control bunker during Operation DESERT STORM, a legitimate military target, is widely viewed as a colossal failure despite the fact that the unfortunate civilian deaths can be seen as morally acceptable within "double effect" parameters. One might reasonably question whether this incident would have received such prominent treatment in earlier, pre- "instant media" conflicts

Kenneth Vaux is concerned that "just as a missile guided down to a target on a laser beam cannot miss, the new warfare can involve no sport or honor."[87] Others, such as Karsten Struhl and George Hunsinger, seek to extend the timeframe and scope to which the principles apply such that low-lethality attacks on infrastructure, resulting in illness and death years down the road, can be deemed in violation of criteria designed to address the immediacy of war, a theme that will be returned to later in this study.[88] Critiques such as these aside, the debate has largely shifted

84 James T. Johnson, *The War To Oust Saddam Hussein: Just War and the New Face of Conflict* (New York: Rowman & Littlefield, 2005), 20–21: emphasis added.

85 Peter Steinfels, "Beliefs; In the Heat of War, Confronting a Moral Principle of Keeping Civilians Immune from Direct Attack," *New York Times* [New York] 6 April 2002, Late, B: 6.

86 Andrew J. Bacevich, "Morality and High Technology," *National Interest* 45 (Fall 1996): 43.

87 Kenneth L. Vaux, *Ethics and the Gulf War* (San Francisco: Westview Press, 1992), xiii.

88 Karsten J. Struhl, "On Just War, Proportionality, and Bombing Civilians," *Radical Philosophy Review* 20.1 (1999): 19; George Hunsinger, "Iraq: Don't Go There," *Christian Century* 119.17 (August 2002): 11.

back to the *jus ad bellum* arena as evidenced by the spirited international debates regarding the moral correctness of commencing military action in Iraq in 1990–91 and again during the run-up to the 2003 invasion.[89]

Non-western Religious and Cultural Reflections

While just war considerations have deeply influenced Western thought it would be incorrect to assume that non-Western cultures have been silent on the subject. Ethical, cultural, and religious limitations in warfare have been evident for much longer than is perhaps generally recognized. Indeed, just war-like language, in the *jus ad bellum*, *jus in bello*, and *jus post bellum* realms, can be found in any number of sources from antiquity forward and from cultures around the world, demonstrating convincingly that critical questions concerning warfare are of concern to all humankind.

Paul Christopher has identified a number of ancient examples of such limitations. In the Far East, China long recognized rules such as the stipulation that "no war should begin without just cause; that the enemy be notified of pending attacks; that no injury be done to the wounded; and that persons and property of innocents be respected."[90] In, *The Art of War*, Sun Tzu outlines a number of prohibitions such as never "strik[ing] elderly men or further injur[ing] an enemy previously wounded" and directs that the army should "treat the captives well, and care for them."[91] Laotse, founder of the Tao religion, wrote "a good general effects his purpose and stops. ... Effects his purpose as a regrettable necessity. Effects his purpose but does not love violence."[92] Clear parallels to both *jus ad bellum* and *jus in bello* criteria are evident.

In 690 B.C., the Babylonian leader Sennacherib differentiated between Jewish leaders, soldiers, and ordinary citizens in a manner quite consistent with Orend's fourth *jus post bellum* principle:

> I assaulted Ekron and killed the officials and patricians who had committed the crime and hung their bodies on poles surrounding the city. The [common] citizens who were guilty of minor crimes I considered prisoners of war. The rest of them, those who were not accused of crimes and misbehavior, I released.[93]

---

89 Johnson and Weigel, *Just War and the Gulf War*, 8–30.; Bishop Wilton D. Gregory, Washington, D.C., United States Council of Catholic Bishops, to President George W. Bush, 13 September 2002.; George W. Bush, President, "President Bush Outlines Iraqi Threat" (Cincinnati, OH: Cincinnati Museum Center, 7 October 2002).

90 Christopher, *Ethics of War and Peace*, 8.

91 Tzu, *Art of War*, 30, 76. Of course, those captives would then join his side.

92 Quoted in Christopher, *Ethics of War and Peace*, 8.

93 Ibid., 9.

The Hindu civilization also produced a number of rules regarding conduct in war that contain elements of both *jus in bello* criteria and limits on conduct found in the Law of Armed Conflict. As outlined in the *Book of Manu*:

> When the king fights with his foes in battle, let him not strike with weapons concealed in wood, nor with such as are barbed, poisoned, or the points of which are blazing with fire. Let him not strike one who ... joins the palms of his hands in supplication, ... nor one who sits down, nor one who says "I am thine"; nor one who sleeps, ... nor one who is disarmed, nor one who looks on without taking part in the fight, ... nor one who has been grievously wounded, ... but in all cases let him remember the duty of honorable warriors.[94]

Similarly, the Lateran Council of 1139 outlawed the use of crossbows against Christians in Europe.[95]

While examining the topic from the perspective of Eastern Orthodox Christianity, Fr. Alexander F.C. Webster offers a key nuance, one worthy of further exploration. To this point, the expression "just war" has been central to the discussion. The *Oxford English Dictionary* (*OED*) defines "just" as, "in the required or appropriate manner."[96] Central to this definition is a sense of righteousness, that a just course is the correct course—something we should do. Webster, however, uses the term "justifiable war," first proposed by the late Paul Ramsey. The term "justifiable" is defined by the *OED* as, "capable of being legally or morally justified; defensible."[97] Here the underlying sentiment seems far different.

Rather than a "just war" being the correct thing to do, a "justifiable war" becomes something that you can morally defend *if*, and this qualifier is critical, you must embark upon that course of action. A sense of "lesser evils" from a selection of bad options seems implied. While acknowledging that there is an ongoing "frontal assault" from some contemporary Eastern Orthodox theologians, Webster outlines six textual sources, from biblical times forward, which underpin the Orthodox moral tradition, noting that, justifiable war enjoys "an unbroken continuity from its origins in Old Testament Israel."[98]

94 Ibid.

95 Martin Van Creveld, *Technology and War: From 2000 B.C. to the Present* (New York: The Free Press, 1991), 71; Eugene B. Skolnikoff, *The Elusive Transformation: Science, Technology, and the Evolution of International Politics* (Princeton, NJ: Princeton University Press, 1993), 51–2.

96 *Oxford English Dictionary*, 2nd ed. (1989), s.v. "just."

97 Ibid., s.v. "justifiable."

98 Fr. Alexander F.C. Webster, "Justifiable War in Eastern Orthodox Christianity," *Just War in Comparative Perspective*, ed. Paul Robinson (Burlington, VT: Ashgate, 2003), 40.

As a faith tradition, "Islam (as Christianity) is characterized by the acceptance of certain practices conceived as legitimate ways of ordering human existence."[99] Given ongoing contemporary world events, how Islam has viewed justice in war during its history might also be illustrative. However, any attempt at obtaining an easy comparison between the Western just war concept of *jus ad bellum* and equivalent Islamic thought, *jihad*, immediately encounters a significant hurdle.

In *Just War Against Terror: The Burden of American Power in a Violent World*, Jean Bethke Elshtain describes the interaction between Western and Muslim scholars at a 1993 conference, "The Ethics of War and Peace," held in Israel. There Bassam Tibi said:

> [The] Western distinction between just and unjust wars linked to specific grounds for war is unknown in Islam. Any war against unbelievers, whatever its immediate ground, is morally justified. Only in this sense can one distinguish just and unjust wars in Islamic tradition. When Muslims wage wars for the dissemination of Islam, it is a just war. ... When non-Muslims attack Muslims, it is an unjust war. *The usual Western interpretation of jihad as a "just war" in the Western sense is, therefore, a misreading of this Islamic concept.*[100]

This short passage reveals that to gain a proper and useful understanding of the commonalities and divergences between western and Muslim thought on the subject, with the necessary nuances and interpretations between earlier and contemporary scholars, would require an effort far beyond the scope of this work.

Similarly, while a quick comparison between the two worldviews in terms of the *jus in bello* branch of the Just War Tradition reveals some commonality, divergent views exist. Frank Trombley, for example, notes that more than a thousand years ago there were clearly identifiable discrimination-like features embedded in the Muslim law of war such as "the immunity granted to Christian monks, as long as they did not take on the role of combatants."[101] Others, however, argue that these moral constraints in war are based less on justice and more on the concept of personal honor.[102]

99 John Kelsay, "Implications of Islamic Political Thought," in *Explorations in Global Ethics: Comparative Religious Ethics and Interreligious Dialogue*, ed. Sumner B. and Bruce Grelle Twiss (Boulder, CO: Westview Press, 2000), 218.

100 Jean Bethke Elshtain, *Just War Against Terror: The Burden of American Power in a Violent World* (New York: Basic Books, 2003), 131.

101 Frank R. Trombley, "The Arabs, the Byzantine State and the Islamic Law of War (*fiqh al-Jihad*) (7th–10th Centuries C.E.)," *Just War in Comparative Perspective*, ed. Paul Robinson (Burlington, VT: Ashgate, 2003), 164.

102 Elshtain, *Just War Against Terror*, 131.

International Laws and Conventions

Hugo Grotius's efforts to translate the essence of Christian just war-like principles into a body of secular work, based on natural laws equally relevant for all humankind regardless of religion, race, or nationality, has rightly earned him the title of "Father of International Law."[103] Today, international laws, the Law of Armed Conflict (LOAC), and related conventions, also seek to limit when nations may legitimately use violence to solve inter-state problems and what measures are permissible in the conduct of war.[104] Not surprisingly, the language used is very similar to that used in just war writings. As Geoffrey Best observes, "It must never be forgotten that the law of war, wherever it began at all, began mainly as a matter of religion and ethics. ... It began in ethics and it has kept one foot in ethics ever since."[105]

While a full examination of applicable laws and conventions is beyond the scope of this study, a few examples here, expanded upon later, are sufficient to show the commonality of thought between the secular legal domain and the ethical, religious, and cultural traditions already examined. The Charter of the United Nations, for example, has language that is congruent with *jus ad bellum* criteria and presumptions against the use of violence. Article 1 declares that the prime purpose of the UN is:

> To maintain international peace and security, and to that end: to take effective collective measures for the prevention and removal of threats to the peace, and for the suppression of acts of aggression or other breaches of the peace, and to bring about by peaceful means, and in conformity with the principles of justice and international law, adjustment or settlement of international disputes or situations which might lead to a breach of the peace.[106]

Article 42 reflects legitimate authority considerations stating that only the Security Council "may take such action by air, sea, or land forces as may be necessary to maintain or restore international peace and security."[107]

In an example from a regional charter, The North Atlantic Treaty intentionally aligns itself with the UN. In Article 1:

---

103 For Grotius' definition of "natural law," see Christopher, *Ethics of War and Peace*, 68.

104 More problematic is the issue of *intra*-state violence and how to deal with non-state actors.

105 Charles J. Dunlap, Jr., "Technology: Recomplicating Moral Life for the Nation's Defenders," *Parameters* XXIX, no. 3 (Autumn 1999): 25.

106 The UN Charter is at: http://www.un.org/aboutun/charter.

107 Ibid.

> The Parties undertake, as set forth in the Charter of the United Nations, to settle any international dispute in which they may be involved by peaceful means in such a manner that international peace and security and justice are not endangered, and to refrain in their international relations from the threat or use of force in any manner inconsistent with the purposes of the United Nations.[108]

Just cause and right intentions considerations are addressed by Article 5:

> The Parties agree that an armed attack against one or more of them in Europe or North America shall be considered an attack against them all and consequently they agree that, if such an armed attack occurs, each of them, in exercise of the right of individual or collective self-defence recognised by Article 51 of the Charter of the United Nations, will assist the Party or Parties so attacked by taking forthwith, individually and in concert with the other Parties, such action as it deems necessary, including the use of armed force, to restore and maintain the security of the North Atlantic area.
>
> Any such armed attack and all measures taken as a result thereof shall immediately be reported to the Security Council. Such measures shall be terminated when the Security Council has taken the measures necessary to restore and maintain international peace and security.[109]

Jus in bello-like language has more commonly been associated with the Hague and Geneva Conventions, the "Hague Law [being] concerned mainly with the means and methods of warfare, while Geneva Law is concerned with protecting persons involved in conflicts, such as POWs, the wounded, and civilians."[110] Article 28 of Hague II (1899) addresses proportionality stating "The pillage of a town or place, even when taken by assault, is prohibited," while Declaration II addressed discrimination concerns by outlawing the "use of projectiles the object of which is the diffusion of asphyxiating or deleterious gases."[111] Similarly, Geneva Convention III (1949) deals with the protection of civilians in the time of war.[112]

Returning to the two questions posed earlier in this chapter, "How then should we respond to this moral quandary, and what role does technology play in that response?" we can now reasonably address the first. The answer that emerges from

---

108 *The NATO Handbook: 50th Anniversary Edition* (Brussels: Office of Information and Press, 1998), 395.

109 Ibid., 396.

110 Dunlap, "Recomplicating," n 16.

111 Article 28 can be found at http://www.yale.edu/lawweb/avalon/lawofwar/hague02.htm while Declaration II can be found at http://www.yale.edu/lawweb/avalon/lawofwar/dec99-02.htm (accessed 1 November 2006).

112 Text can be found at http://www.yale.edu/lawweb/avalon/lawofwar/geneva07.htm (accessed 1 November 2006).

the preceding discussion of various mitigation frameworks, across time, religions, and cultures, is that *humankind should seek to use violence in as few instances as possible, attempt to limit human suffering to the greatest extent possible during war, and seek to establish a better peace after the conflict ends*. But what is the role of technology in this process?

Chapter 2
Technology in War: An Evolving Role But With Voices Raised in Warning

> So by the benefit of this light of reason, they have found out artillery, by which wars come to a quicker end than heretofore, and the great expense of blood is avoided; for the numbers slain now, since the invention of artillery, are much less than before, when the sword was the executioner.
>
> John Donne, 1621

> Weapons, if the right ones can be found, constitute ninety percent of victory.
>
> J.F.C. Fuller, circa World War I

Clearly technology has granted us the power to do much good. In the medical field, advances such as the heart-lung machine, kidney dialysis, CAT scans, and MRIs, are widely perceived as "good" technologies, aiding in the maintenance of life and the diagnosis of disease. Other medical advances push the edge of the "comfort zone" for many. Whereas the practice of *in vitro* fertilization has been widely embraced, abortion, genetic engineering, and cloning alarm many. Similarly, as has been played out on the national stage, end of life issues remain emotionally difficult and a heated religious and political battleground.[1] A similar dichotomy for technologies related to war exists as well.

The hope has long been, as evidenced by the two epigraphs printed above and written some three hundred years apart, that technology would provide humankind a ready means for engaging in quick, and relatively bloodless, conflict. However, the reality seems to be that advances in technology have been, for most of history, enablers of ever-increasing levels of violence. Today, the technology that appears to promise lower levels of violence in warfare is the PGM—perhaps this time the hope will be realized. However, as Edward Tenner warns, "things bite back," unintended consequences often intrude in unexpected and frustrating ways.[2]

Tenner is not alone in warning of technology's potentially insidious influence on humankind. Other voices, such as Capra, Postman, Mitcham, and Ellul, approaching from a number of philosophical perspectives, caution us against blindly following technology's lead. Others, Feenberg among them, offer a perhaps more moderate view. Following an exploration of technology's evolving role in warfare, we will examine these concerns.

---

1 The Terry Schiavo case as one example.

2 See Tenner's, *Why Things Bite Back: Technology and the Revenge of Unintended Consequences* (New York: Alfred A. Knopf, 1996).

Technology's Evolving Role in the Use of Violence

Before examining technology's role in the use of violence in war, it is useful to understand the word itself. Etymologically, the word evolved from the Greek roots techne (*τέχνη*), meaning art, craft, or skill, and logia (*λογία*), meaning study or knowledge. Combined, technology becomes: "a discourse or treatise on an art or arts, or; the scientific study of the practical or industrial arts."[3] This relatively simple definition, based in antiquity, has proven inadequate for many. There are a number of reasons for this. Perspectives will vary depending on whether technology is seen as an object, a process, knowledge, or volition. Definitions are also influenced by the author's personal view of technology as a liberator, a threat, or merely an instrument of power.[4] Carl Mitcham argues that two historical traditions account for many of the various philosophical differences.

The older, what he calls the "engineering philosophy of technology," is perhaps best characterized by its two antecedents, "mechanical philosophy" and "philosophy of manufactures."[5] The engineering philosophy of technology analyzes "technology from within, ... oriented toward an understanding of the technological way of being-in-the-world as paradigmatic for other kinds of thought and action."[6] The second tradition, the "humanities philosophy of technology," is "the attempt of religion, poetry, and philosophy to bring non- or trans-technological perspectives to bear on interpreting the meanings of technology."[7]

The former perspective yields definitions ranging from technology as "applied science" to Peter Engelmeier's "the spring of the great world clock of human development" and "the inner idea of all purposeful action, grounded in the anthropological value of a technological will, which springs from the utilitarian drives" to a contemporary interpretation of technology as "the products of modern engineering."[8] The latter perspective yields definitions of technology such as Frederick Ferré's "practical implementation of knowledge" and Ernst Jünger's "the mobilization of the world through the *Gestalt* of the worker."[9] Mitcham wisely advises that the "tension between the narrow engineering usage and the broad social science usage of the word 'technology' cannot be neatly resolved; it can only be accommodated."[10] Heeding his advice, for this study John Kenneth Galbraith's

---

3 *Oxford English Dictionary*, 2nd ed. (1989), s.v. "technology."

4 Ian G. Barbour, *Ethics in an Age of Technology: The Gifford Lectures Volume Two* (San Francisco: HarperSanFrancisco, 1993), 3.

5 Mitcham, *Thinking Through*, 19.

6 Ibid., 39.

7 Ibid.

8 Ibid., 27, 161.

9 Ibid., 156, 249.

10 Ibid., 151.

definition of technology as the "systematic application of science or other organized knowledge to practical tasks" will be adopted.[11]

Against this backdrop, it can be seen that for thousands of years technological advances, both mechanical and organizational, served primarily as enablers of ever-*increasing* levels of violence. If war was to compel your adversary to do your will, as Clausewitz suggested, traditionally that meant causing enough death and destruction to the enemy's forces to force capitulation on terms favorable to you. While an over-simplification, since other factors clearly entered into the decision-making process of opposing leaders, the underlying point remains valid. The shield, sword, and spear of Greek hoplite warfare eventually evolved into the highly disciplined, tight-ranked musket lines of European infantry of the eighteenth century.

The increased use of focused artillery from the Napoleonic era onward resulted in a more dispersed and entrenched battlefield as seen in World War I.[12] However, possessing the newest and best technology was not, in and of itself, a panacea. As Skolnikoff points out, "superior technology alone by no means guarantees its effective use or actual military superiority. The introduction of machine guns in World War I heavily favored the defense against massed charges, but military leaders refused to accept the overwhelmingly tragic evidence of the battlefield."[13] Finding it ever more difficult to achieve a decisive victory on the battlefield, strategic leaders in World War II renewed efforts to influence the outcome of war by trying to break the will of civilian populations through direct attack.[14]

The culmination of this "bigger bang" process during the twentieth century was the development of nuclear weapons, considered "the most consequential change in military forces in this century ... particularly the coupling of fusion (hydrogen) bombs with intercontinental ballistic missiles."[15] Yet even as these technologies were maturing, dilemmas emerged. As Brodie explains, from World War I into the Cold War American military thinkers were convinced that "modern war is total war."[16] Nuclear weapons were viewed by many as "war preventers," making the outcome of war so horrific that it would not be attempted in the first place, clearly a positive in terms of reducing human suffering.[17] How then was one to deal with smaller "brush fires?"

Unexpectedly engaged in limited wars in Korea and Vietnam, the U.S. found that nuclear weapons, for reasons that included ethical considerations, had no utility

---

11 Ibid., 153.

12 Van Creveld, *Technology and War*, 173.

13 Skolnikoff, *Elusive Transformation*, 50–51.

14 Laying siege to medieval cities and the January 1871 bombardment of Paris during the Franco-Prussian War serve as earlier examples, while the massive air strikes into Hanoi during the Vietnam War serve as later examples.

15 Skolnikoff, *Elusive Transformation*, 51.

16 Bernard Brodie, *War and Politics* (New York: Macmillan Publishing Co., 1973), 63.

17 Not the first time in history a technology was viewed as such—Alfred Nobel's dynamite is another.

in these conflicts.[18] At the same time, countering conventional forces resulted in horrific levels of suffering for both military and civilian populations—without achieving the desired political objectives. Mindful of the post-WWII moral outcry over the deliberate fire bombings of Dresden and Tokyo, echoing the Caledonian Calgacus' observation *solitudinem faciunt, pacem appellant*, perhaps technology, once again, seemed to offer a possible solution, this time in the form of PGMs.[19]

It would be incorrect to state that PGMs first emerged during the Vietnam War. Certainly, the German FX-1400 "Fritz X" and Henschel Hs-239 stand as early attempts to deliver firepower to a relatively precise point from a distance that, ideally, was safe for the personnel employing the weapon.[20] Similarly, the AZON (Azimuth Only), RAZON (Range and Azimuth Only), and TARZON ("Tall Boy" Range and Azimuth Only), the three operational members of the U.S. Army Air Corps/U.S. Air Force VB (vertical bomb) series of glide bombs, represented early U.S. efforts at precision-guidance that did, in fact, experience some limited operational success.[21]

But it would be correct to state that, beginning with the successful introduction of the electro-optically (TV) guided Walleye I glide-bomb into active service in Vietnam in 1967, American military, senior political leaders, and the American populace appear to have developed an increasing fascination with, and reliance on, such weapons.[22] Now guided by electro-optics, infra-red seekers, laser beams, and global positioning satellites (GPS), these weapons have publicly released accuracies of as little as 15–30 feet.[23] The era of placing a weapon precisely at someone's front door appears to be upon us. This potentially war-changing transformation through technology did not escape the notice of other military observers. In the 1980s the Soviets predicted a coming "military-technical" revolution (MTR) enabled by, among other things, "nonnuclear munitions so accurate and lethal that they could wreak levels of military damage comparable to

18 Brodie outlines a number of reasons why nuclear weapons were not used in *War and Politics*, 64–6.

19 Most commonly translated "They make a desert and call it peace."

20 Donald I. Blackwelder, *The Long Road to Desert Storm: The Development of Precision Guided Bombs* (Maxwell AFB, AL: Air University Press, 1993), 2–5. For the roots of precision guidance see Paul G. Gillespie, "Precision Guided Munitions: Constructing a Bomb More Potent Than the A-Bomb" (Ph.D. diss., History, Lehigh University, 2002), 18–120.

21 Ibid., 7–12. At one time, or another, there were five "vertical bombs" under development. These three were used operationally. See also Kenneth P. Werrell, *Chasing the Silver Bullet: U.S. Air Force Weapons Development from Vietnam to Desert Storm* (Washington, DC: Smithsonian Books, 2003), 140–42.

22 Another early Navy PGM, Bullpup, was deployed in Southeast Asia but performed poorly in combat. See Werrell, *Chasing the Silver Bullet*, 142–3.

23 W. Seth Carus, *Cruise Missile Proliferation in the 1990s*, The Washington Papers, ed. The Center for Strategic and International Studies, vol. 159 (Westport, CT: Praeger, 1992), 18; Accuracies for munitions are usually expressed in terms of Circular Error Probable, or CEP, meaning that 50 percent of weapons released will fall within that distance.

those attainable with tactical nuclear weapons," and without the attendant negative effects such as fallout and persistent radiation.[24]

The ability of such weapons to target discrete points on the ground means less explosive power is needed thereby reducing collateral damage to surrounding structures and humans.[25] Such a capability is positive when viewed through the lens of Bentham and Mill's utilitarianism, where the goal is to produce "the greatest overall [beneficial] consequences for everyone."[26] George and Meredith Friedman, in *The Future of War*, give voice to such hopes stating:

> The accuracy of PGMs promises to give us a very different age, perhaps a more humane one. It is odd to speak favorably about the moral character of a weapon, but the image of a Tomahawk missile slamming precisely into its target when contrasted with the strategic bombardments of World War II does in fact contain deep moral message and meaning. War may well be a ubiquitous part of the human condition, but war's permanence does not necessarily mean that the slaughters of the 20th century are permanent.[27]

In a city or other densely populated area, their employment would ideally result in fewer civilian casualties. In the field, fewer military casualties would result as individual pieces of equipment could be designated rather than the targeting of large areas. As PGMs gain greater standoff capability, the ability to be launched at a distance and powered to the aimpoint, fewer of our own forces would also be subject to the lethal defenses of the enemy. Finally, and perhaps the most attractive strategic payoff if indeed true, the ability to destroy discrete targets seen as vitally important by opposing leaders may convince them that further resistance is futile, thereby allowing for a quick, decisive victory with minimal destruction and attendant human suffering. This fundamental concept underpinned the initial *Instant Thunder* air campaign plan for Operation DESERT STORM, the 1991 Coalition effort to eject Iraq from Kuwait.

Technology, in the form of precision-guided munitions may realize the often hoped for potential to act as an enabler of *decreasing* levels of violence in warfare, a fact that has led Philip Meilinger to posit that PGMs should always be "our

24 Thomas A. Keaney and Eliot A. Cohen, *Revolution in Warfare?: Air Power in the Persian Gulf* (Annapolis: Naval Institute Press, 1995), 188. By the mid-1990s some, such as Gen. Yevgeni Volkov, felt that Russia "should enjoy the right to consider the first [enemy] use of precision weapons as the beginning of an unrestricted nuclear war against it." Quoted in, David R. Markow, "The Russians and Their Nukes," *Air Force Magazine*, no. 2 (February 1997): 41.

25 The development of the GBU-39 "Small Diameter Bomb" serves as confirmation.

26 Hinman, *Pluralistic Approach*, 163.

27 Quoted in Dunlap, Jr., "Recomplicating," 24. PGMs do not represent the first case; gunpowder and artillery were seen as such in the seventeenth century.

weapon of first resort."[28] Arguing in clear, ethical language, James Turner Johnson summarizes the case for such weapons stating:

> Given the twin aims of avoiding harm to noncombatants and avoiding more destruction than necessary for a particular justified end, there is a moral obligation to develop weapons and tactics that better serve these ends and to incorporate these weapons into military training, operational plans, and combat decisions by commanders and individual soldiers. This, in moral terms, is the rationale for the development and adoption of precision-guided munitions (PGMs).[29]

Returning then to the second of the two questions posed we can now offer a qualified response. Across time, technology's role has generally been seen as an enabler of ever-increasing levels of violence and suffering in conflict. However, precision-guided munitions *may*, for the first time, mark a paradigm shift, exemplifying technology's emerging role as an enabler of ever-decreasing levels of violence and human suffering in war. But while many are enamored with such technological advancements a number of voices urge caution. These concerns deal with the issue of technology's role in society in general, although specific technologies or artifacts are generally used to demonstrate the wider points of the argument. To fully understand the impact that PGMs are having in contemporary conflicts, and to attempt to remove the qualifier from the answer above, the concerns that have been raised by various thinkers must be both understood and addressed.

Voices Raised in Warning: Concerns Regarding Technology's Impact and Influence on Humankind

With the ever-increasing role that technology has played in society, observed by many as detrimental or worrisome, many philosophers have entered the discussion. Indeed, volumes have been written on the subject of "Philosophy and Technology."[30] In *Ethics in an Age of Technology: The Gifford Lectures Volume Two* Ian Barbour opens with a discussion of three divergent views of technology. As he describes,

> Some see it as the beneficent source of higher living standards, improved health, and better communications. ... Others are critical of technology, holding that it leads to alienation from nature, environmental destruction, the mechanization of

---

28 Phillip S. Meilinger, "Precision Aerospace Power, Discrimination, and the Future of War," *Royal Air Force Air Power Review* 4.3 (Autumn 2001): 25.

29 Johnson, *The War To Oust Saddam Hussein*, 78.

30 See, for example, *Philosophy of Technology: The Technological Condition. An Anthology*, ed. Robert C. Scharff and Val Dusek, Blackwell Philosophy Anthologies (Malden, MA: Blackwell Publishing, 2003).

> human life, and the loss of human freedom. A third group asserts that technology is ambiguous, its impacts varying according to the social context in which it is designed and used, because it is both a product and a source of economic and political power.[31]

The first view is commonly held in society today, a point that is clearly a direct concern for the latter two groups. However, an examination of the opinions of, and tensions between, these two groups of thinkers provides some important insights when examining PGMs.

Technology as Threat: A Danger to Humankind

Fritjof Capra, in *The Turning Point: Science, Society, and the Rising Culture*, argues that the world is in "a crisis of intellectual, moral, and spiritual dimensions … unprecedented in recorded human history."[32] His principal theme is that worldwide problems are but outward manifestations of "a crisis of perception." Because we seek solutions to the problems of a post-modern, interconnected world through the lenses of "Cartesian-Newtonian science," we "face the very real threat of extinction of the human race and of all life on this planet."[33] He believes there is a fundamental problem with our corporate mindset, a firm conviction that we are too narrowly focused on critical issues. While perhaps a bit overplayed, as he is writing at the height of tensions between the two super powers of the Cold War, he nevertheless has a valid point.

As examples of this reductionist approach, he points out that while cardiac surgeons can operate on the heart, they cannot realistically ignore the psychological impacts of the procedure—but many do just that. In the field of psychology, practitioners have "adopted [a] strict division between the *res cogitans* and the *res extensa*" the mind and the body.[34] And, "economists fail to recognize that the economy is merely one aspect of a whole ecological and social fabric."[35] To Capra, all of these technical specialists have lost the wider perspective, a theme of "narrow sightedness" which emerges in other's writings as well. The answer, he believes, is to be found in a fundamental shift in focus from "a mechanistic to [a] holistic conception of reality."[36]

Similarly, Neil Postman argues strongly against the blind acceptance of purely technological solutions. He believes that Sir Charles P. Snow missed the larger point that "the argument is not between humanists and scientists but between

---

31 Barbour, *Ethics in an Age of Technology*, 3.

32 Fritjof Capra, *The Turning Point: Science, Society, and the Rising Culture* (New York: Bantam, 1982), 21.

33 Ibid., 15, 21.

34 Ibid., 164.

35 Ibid., 188.

36 Ibid., 16.

technology and everybody else."[37] While acknowledging technology "is a friend, [making] life easier, cleaner, longer," he warns that, "the uncontrolled growth of technology destroys the vital sources of our humanity. It creates a culture without a moral foundation."[38]

Postman's main concern is that our culture has been far too intimate with technology, and for far too long. As a result, "technology does not invite a close examination of its own consequences. It is the kind of friend that asks for trust and obedience, which most people are inclined to give because its gifts are truly bountiful," a clear articulation of Barbour's first group.[39] To make his case, he explores a cultural shift that he believes has taken place, moving humankind, albeit not all at the same time, from tool-using cultures, to technocracies, to technopolies.[40] Postman defines a technocracy as "a society only loosely controlled by social custom and religious tradition and driven by the impulse to invent."[41] Technopolies, clearly a pejorative term for Postman, are "totalitarian technocracies," societies that deify technology, seeking both their authority from, and finding satisfaction in, technology. Again the cautionary theme of a too narrow, single-minded surrender to technology emerges. Significantly, he identifies America as the first "technopoly."[42]

Having acted as social critic, Postman offers a response, although self-admittedly not a solution. Articulated in two parts, the role of the individual that is most important to this study. In considerable harmony with Capra, he asks that individuals act as "loving resistance fighters," individuals who "admire technological ingenuity but do not think it represents the highest form of human achievement."[43] Such individuals understand that,

> Technology must never be accepted as part of the natural order of things, that every technology ... is a product of a particular economic and political context and carries with it a program, an agenda, and a philosophy that may or may not be life-enhancing and that therefore require scrutiny, criticism, and control. In short, a technological resistance fighter maintains an epistemological and psychic distance from any technology, so that it always appears somewhat strange, never inevitable, never natural.[44]

37 Neil Postman, *Technopoly: The Surrender of Culture to Technology* (New York: Vintage Books, 1992), xii. Snow gave the now famous "Two Cultures" Rede lecture at Cambridge University. His argument that scientists and humanists did not understand each other was viciously attacked by the literary scholar F.R. Leavis. See, Yuri Levin. "Snow's Two Cultures—and Ours." *Public Interest*. No. 153, Fall 2003, 54–69.

38 Ibid.

39 Ibid.

40 Ibid., 22.

41 Ibid., 41.

42 Ibid., 48, 71.

43 Ibid., 184.

44 Ibid., 184–5.

The critiques along this line of reasoning continue. Carl Mitcham considers technology "largely an unthinking activity."[45] Lewis Mumford, seeking to keep our humanity in focus, argued "the machine should be considered in terms of 'its psychological as well as its practical origins' and appraised in aesthetic and ethical terms as much as technical ones."[46] These philosophers clearly hold largely critical views of technology, and are mainly concerned with what they see as the very real possibility of technology taking over and dominating humankind, largely due to our lack of active participation and/or vigilance. One, the French philosopher Jacques Ellul, offers a complex narrative that, while largely capturing the essential concerns of Capra, Postman, Mitcham and others, identifies a specific modality that, he believes, threatens human freedom, and which dovetails neatly with the wider discussion of precision munitions this study has embarked upon—his concept of *technique*, which will be examined shortly.

Technology as Contingent: Humankind "In the Loop"

It must be acknowledged that other philosophers, while not the all accepting technophiles of the first group, are somewhat less strident about the dire consequences of human-technology interaction than the second. For more than three decades, Andrew Feenberg has been a leading American voice in this third of Barbour's groups. Rejecting the notions of the social critics who claim that humankind has, or must, accept the preordained path that modern technology has prepared for it, he similarly distances himself from those who profess that technology must be fully abandoned and that we must return to a simpler, less technical time to save ourselves, and the planet.

In doing so, Feenberg rejects the idea that technology is deterministic in nature, the belief that "technologies have an autonomous functional logic that can be explained without reference to society."[47] Instead, he adopts a more Kuhnian perspective, taking a historicist's view that maintains that technology, because of its social aspect, is a contingent process, which can lead in many directions.[48] Dismissive of the position that humankind has not had a hand in the development effort, Feenberg is most concerned with the fact that relatively few actors actually have a hand in this contingent process. Ultimately, what Feenberg seeks is a change to the current bureaucratic structure where "owners of corporations,

45 Mitcham, *Thinking Through Technology*, 1.

46 Quoted in Ibid., 41.

47 Andrew Feenberg, "Subversive Rationalization: Technology, Power, and Democracy," in *Technology and the Politics of Knowledge*, ed. Andrew Feenberg and Alastair Hannay (Bloomington: Indiana University Press, 1995), 6.

48 A summary of Kuhn's approach is in Andrew Feenberg's *Questioning Technology* (New York: Routledge, 1999), 89–90.

military bureaucrats, and the professional organizations of technologists have far more influence over [the design process] than ordinary citizens."[49]

He acknowledges that when the public deems an issue "important enough" to become involved in, it, or at the very least a dedicated vocal minority, can and will divert a technical object off the path that it would otherwise have taken purely in the hands of the aforementioned elite. Why then, he asks, does this not happen more often?

As opposed to Heidegger, who awaited "the arrival of the new gods needed to open up a new world," Feenberg believes that a new societal structure can re-empower the disenfranchised voice of the workers and the general public in the overall technical design process.[50] This new structure, evolving through a process that he calls "democratic rationalization," would offer the world a new and better relationship with technology.

For years, Feenberg called this process "subversive rationalization." While in his third book, *Questioning Technology*, he changed the name, attributing this action to "a more developed analysis of technical politics," the rationale remained the same.[51] The methodological crux of this process is that "new technology can also be used to undermine the existing social hierarchy or force it to meet needs it has ignored."[52] In essence, the public engages the elite to force the needed structural change.

In its previous, less technical, manifestation this process was at work in union struggles over health issues, struggles in Europe over the deployment of nuclear weapons, environmental protests over logging policies, and community protests over chemical plants and toxic waste. But he sees these movements as disparate parts of what could, nay should, become a coherent whole. The emergence of the Internet and other networking technologies is, he believes, the possible binding agent.

"Networking has given rise to one among many such innovative public reactions to technology. Individuals who are incorporated into new types of technical networks have learned to resist through the net itself in order to influence the powers that control it."[53] Perhaps the networking technology, developed by the elite, has within itself the seeds of its own reform. Thus democratic rationalization can help give birth to the technical democracy—the reform Feenberg so hopes for and sees as humankind's salvation.

49 Andrew Feenberg, *Alternative Modernity: The Technological Turn in Philosophy and Social Theory* (Berkeley: University of California Press, 1995), 5.

50 Michael E. Zimmerman, *Heidegger's Confrontation with Modernity: Technology, Politics, Art* (Bloomington: Indiana University Press, 1990), 92.

51 Andrew Feenberg, *Questioning Technology* (New York: Routledge, 1999), xvi.

52 Andrew Feenberg "Escaping the Iron Cage, or, Subversive Rationalization and Democratic Theory," in *Democratising Technology: Ethics, Risk, and Public Debate*, ed. R. Schomberg (Tilburg: International Centre for Human and Public Affairs, 1998), 1.

53 Ibid., 19.

Ellul and *Technique*

To comprehend the writings of Jacques Ellul, it is necessary to have a basic understanding of the man, the times in which he matured, and the key influences on his life. Born in France in 1912, Ellul's family, once rich, had suffered significant financial setback. As James Fowler notes, Ellul "considered this one of the decisive elements in the development of his life and thinking, that he was acquainted with poverty."[54] He matured during the inter-war period and, as did many others of that era, embraced Marxism in 1931. Just three years later, he experienced his "second conversion" and embraced Christianity.[55] Christian observes that Ellul "lived with this contradiction until his death in 1994. Marx and the Bible were dialectic in his mind, with his sociological pole rooted in Marx and his religious commitments in [the Swiss theologian] Karl Barth."[56]

A lawyer by training, he spent most of his adult life on the faculty of the University of Bordeaux. A short, albeit frustrating, stint in local politics was followed by lifelong involvement with the Church and other social action causes, which evidences, perhaps, an outward manifestation of the aforementioned intellectual contradiction.[57] A prolific author, he saw himself "not a philosopher, not a theologian, not a biblical scholar, not a *specialist*, period" but as one who "is describing a group situation in which he is himself involved."[58] *The Technological Society*, published in English in 1964 and seen by many as his most influential work, attracted a great deal of attention and placed his concept of *technique* firmly into the lexicon of the technology/religion/humanity debate.[59]

Technique: A Complex Abstraction

Jacques Ellul defines *technique* as "the totality of methods rationally arrived at and having *absolute efficiency* (for a given stage of development) in every

---

54 James A. Fowler, "A Synopsis and Analysis of the Thought and Writings of Jacques Ellul," Article, 2000, 18 July 2006 http://www.christinyou.net/pages/ellel.html.

55 Ibid. Also Clifford G. Christians. "Jacques Ellul's Conversions and Protestant Theology." *Journal of Media and Religion* 5.3 (2006).

56 Christians, "Ellul's Conversions," abstract.

57 See Jacques Ellul and Patrick Troude-Chastenet, *On Religion, Technology, and Politics: Conversations with Patrick Troude-Chastenet*. Translated by Joan Mendes France. South Florida-Rochester-Saint Louis Studies on Religion and the Social Order, vol. 24. (Atlanta: Scholars Press, 1998), 15.

58 The first quote is from John Wilson, "Just an Ordinary Jacques," review of *What I Believe*, by Jacques Ellul, *Christianity Today* 35.5 (29 April 1991): 52.; the second is Ellul's own description from Jacques Ellul, *The Technological Society*, trans. John Wilkinson (New York: Vintage Books, 1964), xxvii.

59 It was originally published a decade earlier in France as *La technique ou l'enjeu du siècle* (Technique, the Stake of the Century) without much attention.

field of human activity."[60] It is easily understood how some, particularly those reading in English, could confuse, or simply replace, the word *technique* with the word "technology." Carl Mitcham does just that when introducing Ellul and his philosophy.[61] Yet, as Fowler points out, the two are not the same; not in French, and not in translated English.[62] *Technique* is a complex abstraction, within which technology is but one component part. Ellul himself is clear on this, as he preceded the above definition with the statement, "the term *technique*, as I use it, does not mean machines, technology, or this or that procedure for attaining an end."[63]

Ellul acknowledges early in *The Technological Society* that in ancient times "*technique* certainly began with the machine" but "has now become almost completely independent of [it]."[64] This initially gradual shift, which has accelerated exponentially since the Industrial Revolution, is the key to Ellul's underlying concerns. Writing as a man of faith, he records that the Church questioned early manifestations of *technique*. He explains, "the question 'Is it righteous?' was asked of every attempt to change modes of production or of organization."[65] But, where we once had choice, in modern times the evolution of *technique* into what he calls the "technological phenomenon," has removed that measure of freedom, to our spiritual detriment.[66]

Two factors, reason and consciousness, intervened to produce this phenomenon. Reason permitted, indeed still permits, us to analyze the results of our efforts driving us, over time, to seek the most efficient path to our desired end. Consciousness allows us to recognize the advantages gained by that more efficient path. As Ellul summarizes,

> The two-fold intervention of reason and consciousness in the technical world, which produces the technological phenomenon, can be described as the quest of the one best means in every field. And this "one best means" is, in fact, the technical means. It is the aggregate of these means that produces the technical civilization.[67]

60 Ellul, *The Technological Society*, xxv: emphasis is in the original.
61 Mitcham, *Thinking Through Technology*, 57.
62 Fowler, "Thought and Writings of Jacques Ellul."
63 Ellul, *Technological Society*, xxv.
64 Ibid., 3–4.
65 Ibid., 37.
66 Ibid., 77–9.
67 Ibid., 21.

Technique's Modern Form and its Perceived Dangers

Ellul saw *technique* as a "new and specific milieu in which man is required to exist, one which has supplanted the old milieu of nature."[68] This new milieu has certain characteristics. It is,

> artificial; ... autonomous with respect to values, ideas and the state; ... a closed organization which permits it to be self-determinative independently of all human intervention; ... formed by an accumulation of means which have established primacy over ends; [with] all its parts ... mutually implicated to such a degree that it is impossible to separate them or to settle any technical problem in isolation.[69]

Technique, then, permeates all aspects of our existence. In fact, Ellul argues that humankind has blindly allowed our fascination with technical solutions, to virtually every problem facing us, to embed itself to such a degree that our "goals are represented only by such progress and happiness as is to be achieved through techniques." This process, he concludes, renders man "no longer in his traditional state of freedom with respect to judgment and choice," which effectively shifts man from the subject to the object of *technique.*[70] As a result, the more we embrace this reality, the more "change does not depend on the approval or disapproval of those living in the society."[71] A sense of kinship with Postman's "technopoly" is evident here.

Some interpret this modern relationship with science and technology as having evolved into a new religion, dubbed "techno-secularism," which simultaneously helps supplant old faith traditions, particularly in America and the West.[72] Such thoughts are not alien to Ellul who regarded misplaced faith in technique and technology as the fatal idolatry of the twentieth century."[73] Not surprisingly, given his deep faith, it is a return to foundational Christianity Ellul believed, that might offer humankind some measure of hope. In addition to this loss of human freedom, with its associated social and spiritual consequences, what other aspects of the phenomenon of *technique* concern him?

One critical issue for Ellul is that of unintended consequences. Expanding upon the sentiments articulated above, he is concerned that "technical progress today is no longer conditioned by anything other than its own calculus of efficiency."[74]

68 Jacques Ellul, "Ideas of Technology: The Technological Order," trans. John Wilkinson, *New Perspectives Quarterly* 17.3 (Summer 2000): 21.

69 Ibid.

70 Ibid.

71 Wilson, "Just an Ordinary Jacques," 52.

72 Rustum Roy, "Scientism and Technology as Religions," *Zygon* 40. 4 (December 2005): 835–44.

73 Samuel Hope, "Homage to Jacques Ellul," *Arts Education Policy Review* 97.5 (May/June 1996): 38.

74 Ellul, *Technological Society*, 74.

As a result, in an era when technological *means* become the focus, to the relative exclusion of rationally considering *ends*, "certain unforeseen secondary effects ... are much more disastrous than the lack of technique would have been," citing the pesticide DDT to make his point.[75] Originally thought harmless, when combined with a fatty solution the composition turned out to be poisonous to livestock and children.[76]

Contemporary examples, many in the pharmaceutical field, are also evident.[77] The tensions here are understandable. Imbued with an almost blind fascination with all things technical, consumers want new products as quickly as possible. With high development costs, and long lead times an issue, those who develop such products seek a return on their investment as rapidly as possible. Regulators and "watch dogs," primarily manifest in government agencies, try to ensure the safety of the consumer, often while understaffed and under great pressure from the business community. Under these conditions, in the words of Edward Tenner, "things bite back." Ellul argues that the question here is not of error, he accepts that errors will always occur, but of two factors related to *technique*, "the weight of [which] is such that no obstacle can stop it."[78] In the first place, it is impossible to foresee every possible eventuality to a technical action and, in the second, *technique* demands that everything it produces be brought into the public domain.[79]

Technique's Relevance to this Study

It is the "unintended consequences" aspect of *technique* that is of particular relevance to this study. That the development and continued refinement of PGMs is seeking "absolute efficiency" in the field of human endeavor called war is almost self-evident. Ellul quotes J.U. Nef, who argues, "Technical progress favors war ... because (a) the new weapons have rendered more difficult the distinction between offense and defense; and (b) *they have enormously reduced the pain and anguish implied in the act of killing*."[80] This fact alone would argue for viewing PGMs as a net positive in regards to human suffering. But, as mentioned earlier, it is the examination of Ellul's key concern regarding unintended consequences that is critical to a full examination of the question. If the unintended consequences resulting from the use of PGMs cause an increase in human suffering, exceeding the reduced levels Nef alludes to, then Ellul may be right.

A second manifestation of Ellul's *technique* is also of interest here. In *The Humiliation of the Word*, he examines humankind's increasing fascination with the

75 Ibid., 105.

76 Ibid., 106.

77 The recent VIOXX settlement is one example.

78 Ellul, *Technological Society*, 106.

79 Ibid.

80 Ibid., 100: emphasis added. This was written before PGMs.

"image" to the detriment of the "word." As the volume's translator, Joyce Main Hanks, notes:

> For years, Jacques Ellul has warned repeatedly that our modern addiction to images is a kind of terrorist time bomb ominously ticking away in the comfortable hotel of free democratic society. *The Humiliation of the Word* powerfully and convincingly demonstrates the wholesale abuse of language and this dangerous addiction to images characteristic of modern society. Public officials are "electable" in the United States today only if they project an attractive television image. Reaction to presidential "debates," for example, depends almost entirely on image, not substance, truth, or coherent rational argument.[81]

This aspect of the larger concept is important because much of what is "known" by the public regarding war, has been delivered for years through visual media. One need only think of the nightly reports on the war in Southeast Asia, in the 1960s and 1970s, to the recent and ongoing conflicts in Iraq and Afghanistan today to see this at play. Images become reality, deeper explanation is neither required nor desired, indeed trying to engage in debate can be seen as an attempt at obfuscating that which is clearly true ... "it was on the television!" One challenge for this study will be to move past popular conceptions of PGMs, to examine the facts of the case studies selected.

Can the Influences of Technique be Mastered?

By turns, Ellul seems overly pessimistic and, at times, cautiously optimistic, concerning the question of whether humankind can overcome the phenomenon of *technique*. To Patrick Troude-Chastenet, he once said, "I describe a world with no prospects, but I believe that God accompanies man throughout his whole existence."[82] In his foundational work on *technique* he explained,

> In this study no solution is put forward to the problems raised. Questions are asked, but not answered. ... I do not say that no solutions will be found; I merely aver that in the present social situation there is not even a beginning of a solution, ... Any solutions I might propose would be idealistic and fanciful. In a sense, it would even be dishonest to suggest solutions: the reader might think them real rather than merely literary.[83]

Elsewhere, however, he opens the possibility to some degree of hope.

Rejecting the views of Feenberg and others, he dismisses the notion that humankind, writ large, can master *technique*. "The imperative that man exercise

81 Jacques Ellul, *The Humiliation of the Word*, trans. Joyce Main Hanks (Grand Rapids, MI: William B. Eerdmans Publishing Co., 1985), vii.

82 Ellul and Troude-Chastenet, *Conversations*, 1.

83 Ellul, *Technological Society*, xxxi.

mastery over technical development is facilely accepted by everyone. But factually it simply does not hold."[84] Yet while he believes that the technician, the scientist, and the politician cannot do it, he leaves open the possibility that the individual, for Ellul this meant the Christian, "can doubtless seek the soundest attitude to dominate the techniques at his disposal." Nevertheless, even here, pessimism remains. He continues, "the individual's efforts are powerless to resolve in any way the technical problem in its universality; to accomplish this would mean that all men adopt the same values and the same behavior."[85]

The potential outcome of the interaction between humanity and technology is, in its widest context, an interesting debate to be sure. Some philosophers, Capra and Postman among them, are clearly concerned that humans have willingly allowed technological progress to narrow their focus in general, in the name of specialization, and to assume a largely passive role as technology expands its influence on contemporary society. This group warns humanity to open their eyes to the role technology is playing in their society, to adopt a more holistic approach, and to become involved, before it is too late. Others, such as Feenberg, are more measured, believing that human interaction has never left the process, but agree that more individuals need to be actively involved in the debate.

Ellul, with his fear that a blind adoption of *technique* is driving humans to seek ever greater efficiency to the exclusion of a consideration of its wider effects, emerges as the philosopher whose work most closely addresses the core issues regarding the main technological focus of this study, namely precision guided munitions. His concerns will form a key component of the philosophical underpinning of this study's evaluative framework. The next step is to examine a number of ethical traditions that, when combined with Ellul and the core principles of the Just War Tradition, serve to complete the framework used to evaluate the case studies.

84 Ellul, *Ideas of Technology*, 21.

85 Ibid., 21–3.

Chapter 3
Measuring the Impact: A Comprehensive Framework

Slay them all. God will know His own.

Abbott Arnold Amaury

Those unfamiliar with the martial profession would be forgiven for assuming that all those who bear arms hold beliefs consistent with Abbott Amaury's declaration.[1] True, history is replete with examples of wholesale slaughter committed in the name of one religion, or supposed "just" cause, or another. However, as was outlined previously, throughout history humankind has sought to mitigate the resort to war, and the use of violence in war largely based on ethical or moral considerations and/or traditions.

To be viable, any fully developed ethical tradition should be complete, comprehensive, non-contradictory, and livable. The Just War Tradition is such a tradition. A long line of religious and secular theologians has developed this Western approach to issues related to war over many centuries, a process that continues today with the emerging *jus post bellum* branch.

Nevertheless, this tradition does not sit by itself outside wider Western ethical thought. It is clear that within the Just War Tradition the deeper roots of other, perhaps now more distinct, ethical traditions resonate. Thus, in developing an evaluative framework, it is also helpful to return to these other approaches to explore and describe them. This effort is not intended to veer away from just war, or to create a new ethical viewpoint or new meta-theory, but there are at least three good reasons to examine these other traditions more deeply.

The first is to gain a greater understanding of the core of the Just War Tradition. For example, the *jus ad bellum* principles of just cause and right intentions have clear linkages to the concept of duty. Examining Kant's writings in greater detail should offer additional insights. Similarly, elements of natural law and utilitarianism are present. While Kant, Bentham, and Mill did not invent the traditions they explored, they did highlight and emphasized them in new systematic ways. Next, is the issue of metrics. To be effective and consistent, some method of measuring human suffering must be developed. Examining the more modern articulations of some of these Western ethical roots should reveal valuable tools that will aid in this effort.

1 The epigraph is reportedly the Abbott's response to a Crusader's question on how they were to tell Catholics from Cathars during the sack of Beziers (22 July 1209). See "Massacre of the Pure," *Time*, 28 April 1961, 54.

Finally, and admittedly a fine line is being drawn here, the explicit choice to explore the resonance between the roots of just war and other traditions seeks to make the framework more readily accessible and understandable to a larger cross-section of readers. Earlier it was noted that a core of just war-like reasoning can be detected in non-Western traditions. Neutrality is not sought here, but if properly constructed the effort might avoid the criticism "This framework is no good, it is only based on just war legalism." If others can identify with the language or concepts used, dialogue may continue.

One complicating factor, however, is that there are many moral and ethical frameworks, often at odds with each other in some critical fashion, in use in the world today. A second revolves around the terms "morals" and "ethics" themselves. The *Oxford English Dictionary* defines morals, when used as a noun, as "thought and discourse about moral questions; moral philosophy, ethics" and "a set of personal standards relating to right and wrong conduct." When used as an adjective, "of or relating to human character or behaviour considered as good or bad; of or relating to the distinction between right and wrong, or good and evil, in relation to the actions, desires, or character of responsible human beings; ethical."[2]

Ethics, on the other hand, is defined as "relating to morals," "the moral principles by which a person is guided," "the rules of conduct recognized in certain associations or departments of human life," and "the science of morals; the department of study concerned with the principles of human duty."[3] The confusing admixture of terms is quite apparent from these definitions. In the end, rather than attempting to draw a clear distinction between the two, as some such as Lawrence Hinman have done, it is perhaps best simply to acknowledge that the terms are often used in an interchangeable fashion, even by ethicists.[4]

A Survey of Ethical Systems

The philosophical underpinnings of moral belief systems emerge from a number of perspectives, both intellectual and spiritual. Understanding the core tenets, and critical nuances, of these numerous belief systems is a complex and time-consuming undertaking. One way to simplify such a daunting task is to categorize moral points of view into broad groupings, such as freedom-based, rule-based, or character-based systems.

It is acknowledged that the boundaries between these groupings are sometimes porous, and the various viewpoints will, naturally, overlap at points.[5] While a

2 *Oxford English Dictionary*, 2nd ed. (1989), s.v. "moral."

3 Ibid., s.v. "ethic."

4 See Hinman, *Pluralistic Approach*, 5.

5 Ethicists have devised many rational groupings. This particular set was developed at the U.S. Naval Academy as an in-house guide to educate instructors assigned to teach the introductory ethics course.

detailed examination of the many traditions contained in each of these categories is beyond the scope of this work, expanding upon a few will serve to demonstrate the fundamental characteristics of each, and to help determine if a given tradition can be usefully employed in expanding the understanding of just war and in developing an effective evaluative framework.[6] As freedom-based systems such as anarchy, nihilism, relativism, and egoism are primarily centered on "self," or seek to do without government, they will not be examined.

Rule-based Systems

Rule-based ethical systems, as the name suggests, admit of a set of truths or moral standards that exist to guide proper human behavior. In some traditions, these truths are deemed absolute, true at all times, in all places, for all people. In others, the rules are conditional, serving as guidelines, but seeking the best outcomes possible. Such systems may be rooted in religion, known as the ethics of "divine command," natural law, or in a variety of philosophical viewpoints first brought forth by prominent thinkers during the Age of Reason and the Age of Enlightenment.

Divine command Trending towards the absolutist end of the spectrum, particularly in their stronger forms, are those traditions that ascribe to divine command as their moral compass. Accepting that absolute truths and moral standards come from God, or through divine revelation, the belief is that "*whatever God commands is what is right simply because God commands it.*"[7] Divine law also addresses the question of why we should choose to do the right thing, as it presupposes that a final judgment awaits us all. We should act correctly because God knows what we have done, even if other men do not.

Such theories work best within the world's monotheistic religions, Judaism, Christianity, and Islam, where God is seen as universally good, but fit less well in other world views.[8] However, when determining whether any divinely based tradition can be usefully adopted into the evaluative framework, some obvious problems emerge. First, and implying no criticism of the tenets of the three major monotheistic religions or the adherence to the same by the billions of the faithful, the divine commands contained within these religions are simply not held to be true and valid by all of humankind.

Hinman observes that, such theories "have been subjected to strong criticisms from both theistic and atheistic philosophers who maintain that God's commands cannot override an ethics based on reason."[9] There are a number of grounds for

6 Hinman's work is an excellent primer.

7 Ibid., 97: emphasis in original.

8 Ibid., 98–9. For example, such theories have little to no validity in the worldviews of the Navajo.

9 Ibid., 99.

such criticism, not the least of which is the difficulty of validating that any one human has accurately divined the will of God to the satisfaction of all. The Swiss Reformed theologian Karl Barth argued, "that the human being is constitutionally incapable of coming to know God through any 'natural' human activity—rational, affective, or ethical. God is knowable only insofar as God freely chooses to make Godself known."[10]

This point, coupled with the realization that those holding strong absolutist views grounded in any particular religion are unlikely ever to accept as valid the tenets of another, seems to argue against the adoption of portions of divine command ethics as part of the framework. As Twiss and Grelle have ably demonstrated, agreeing upon a global ethic, while a worthy undertaking, is a daunting if not impossible task. However, one aspect of the ethics of divine command overlaps with the ethics of natural law and that aspect merits further examination.

Natural law With deep roots in early Greek and Roman thinking, the concept of natural law posits that truth and right are part of the natural order of things, and that human consciences are a reliable guide to making right choices. It does not take much examination, however, to identify inequities between peoples in most societies today. In one of his early works, "Justice as Fairness," the American philosopher John Rawls acknowledges these inherent inequalities, noting that:

> men born into different positions have different expectations of life determined, in part, by the political system as well as by economic and social circumstances. In this way the institutions of society favor certain starting positions over others. These are especially deep inequalities. Not only are they pervasive, but they affect men's initial chances in life.[11]

Those in advantageous positions would seem inclined to maintain such advantages, and their choices would reflect, at the very least subconsciously, that reality. How, then, might these inherent inequalities be overcome, and "right choices" made?

In answer, Rawls offers his concept of "original position" and its associated "veil of ignorance." In this purely theoretical starting point, "no one knows his place in society, his class position or social status, nor does anyone know his fortune in the distribution of natural assets and abilities, his intelligence, strength, and the like."[12] Thus positioned, the resultant agreed upon "principals of justice," the "right choices," are made fairly. Rawls believes that from such a starting point two principles of justices would emerge.

10 Sumner B. Twiss and Bruce Grelle, *Explorations in Global Ethics: Comparative Religious Ethics and Interreligious Dialogue* (Boulder, CO: Westview Press, 2000), 76.

11 John [B.] Rawls, *A Theory of Justice* (Cambridge, MA: Belknap Press, 1971), 7.

12 Ibid., 12.

The first is that "each person is to have equal right to the most extensive basic liberty compatible with a similar liberty for others."[13] He defines these basic liberties as "political liberty ... freedom of speech and assembly; liberty of conscience and freedom of thought; freedom of the person along with the right to hold (personal) property; and freedom from arbitrary arrest and seizure as defined by the concept of the rule of law."[14]

The second, two-part principle is more nuanced and requires much greater explanation on Rawls' part. Broadly summarized it posits that while the basic starting position in life will not be level for each individual, there would be a fair equality of opportunity for all and that the inequalities that did exist in society would benefit the most disadvantaged.[15] These principles were developed within the framework of what Rawls describes as a "well-ordered society." A key phrase in his explanation of such a society is "everyone accepts and knows that the others accept the same principles of justice."[16] Here a sense of homogeneity emerges that may be problematic. Can a more diverse society be ordered on the principles of natural law that Rawls has explained theoretically? The answer to this question, at least in one instance, is perhaps.

The Founding Fathers of the United States, a religiously diverse group committed to the concept of religious freedom, based the Declaration of Independence on the principles of natural law.[17] The first sentence makes explicit reference to the "Laws of Nature" while the second declares, "We hold these truths to be self-evident, that all men are created equal, that they are endowed by their Creator with certain unalienable Rights, that among these are Life, Liberty and the pursuit of Happiness."[18] Writing later in his life, Jefferson observed, "The equal rights of man and the happiness of every individual are now acknowledged to be the only legitimate objects of government."[19] Much common ground with Rawls is evident here.

However, an intertwining of the ethics of natural law and that of divine command, also evident here, occurs because of the further belief, held by many including the Roman Catholic Church, that God created the natural order. The logic develops as follows:

> Anything that comes from an all-good source (i.e., God) must be good insofar as it is true to its origins. Furthermore, in many religious traditions, the natural

13 Ibid., 60.

14 Ibid., 61.

15 Ibid., 60–67.

16 Ibid., 5. Expanded explanation at 453–62.

17 The American system of Constitutional Law is also based on these principles.

18 Quoted in *The Political Writings of Thomas Jefferson: Representative Selections*, ed. Edward Dumbauld, Reprinted, The American Heritage Series, vol. No. 9 (New York: The Liberal Arts Press, 1956), 3.

19 Ibid., xxv. Originally in a letter written to A. Coray on 31 October 1823.

> order is oriented towards a final purpose or goal, union with the divine, which is necessarily good. *After the goodness of the natural order is established in this way, it is a short step to claiming that people are entitled (that is, have a right) to whatever fulfills the natural order.*[20]

Clearly, not all adherents to the concept of natural law emerge from such a theistic foundation, and this is a significant point. It must also be acknowledged that the strong political commonality of purpose for the Founding Fathers might have been strong enough to overcome other divergent views thus making them resemble a "well-ordered society." But if portions of the ethics of natural law, and its subset natural rights, appeal to the religiously inclined, to those who are not, and to people of diverse cultural backgrounds, then elements of this tradition could usefully be included in the evaluative framework precisely because they bridge this difficult religious-secular-cultural divide. Can such a consensus be found?

In a later work, *Political Liberalism*, John Rawls, reworks his earlier writings and addresses the diversity of modern democratic societies. Accepting that "there is no reasonable religious, philosophical, or moral doctrine affirmed by all citizens, a concept of justice affirmed in a well-ordered democratic society must be a conception limited to what I call the 'domain of the political.'"[21] By achieving this political conception of justice, Rawls hopes to gain the support of an "overlapping consensus." Such a consensus "consists of all the reasonable opposing religious, philosophical, and moral doctrines likely to persist over generations and to gain a sizeable body of adherents in a more or less constitutional regime, a regime in which the criterion of justice is that conception itself."[22]

Rawls argues that concept of overlapping consensus can be misunderstood, and that society should not merely seek to find core commonalities in existing doctrines which, when combined, might obtain the desired outcome.[23] In practice, however, when expanding the focus from single societies to transnational bodies it appears that a number of cross-cultural documents in place today, that use natural law as their basis, were achieved in such a fashion and have managed to gain a fair degree of consensus among widely diverse groups of people. For example, the preamble to the North Atlantic Treaty uses natural rights language declaring, members are determined "to safeguard the freedom ... of their peoples, founded on the principles of democracy, *individual liberty* and the rule of law."[24]

Moving beyond this primarily western European context, the United Nations Charter and the organization's Universal Declaration of Human Rights are also

20 Hinman, *Pluralistic Approach*, 254: emphasis mine.

21 John [B] Rawls, *Political Liberalism*, (New York: Columbia University Press, 1993), 38.

22 Ibid., 15.

23 Ibid., 39.

24 *The NATO Handbook: 50th Anniversary Edition* (Brussels: Office of Information and Press, 1998), 395: emphasis mine.

grounded in the concepts of natural law/natural rights. The preamble to the Charter begins, "We the people of the United Nations determined ... to reaffirm faith in fundamental human rights, in the dignity and worth of the human person, in the equal rights of men and women and of nations large and small, and ... to promote social progress and better standards of life in larger freedom."[25] While the issue of human rights is touched upon again in Articles 1, 55, and 76 of the Charter, a full explication of what that term means in practice was not to come until the adoption by the General Assembly of the *Universal Declaration of Human Rights* on 10 December 1948. The preamble to the latter document speaks of the "recognition of the inherent dignity and of the equal and inalienable rights of all members of the human family," the main body continues with statements that are in clear agreement with the sentiments of Jefferson almost two centuries earlier.[26] Article 1, for example, states, "All human beings are born free and equal in dignity and rights," while Article 3 declares "Everyone has the right to life, liberty and security of person."[27] The remainder of the document expands upon these broad themes, adding much greater specificity.

Despite the fact that this document resulted from an extensive, collaborative effort among many diverse cultures, some have questioned whether this document, and indeed others like it, merely seeks to impose a Western set of ideals and morals regarding human rights on the whole of humankind.[28] But ample evidence exists demonstrating that the document enjoys wide support in areas as culturally diverse as Africa, Asia, and the sub-continent.[29] In fact, Darrell Fasching believes, "The covenant embodied in the UN declaration is at once both holy and secular. It cuts across the religious and the secular, winning adherents both religious and nonreligious."[30] Returning then to a view expressed earlier, it is for just this reason that considerations emerging from the natural laws-natural rights school, using

25 Charter of the United Nations (1945). Available at: http://www.un.org/aboutun/charter/ (accessed 28 February 2008).

26 Universal Declaration of Human Rights (1948). Available at: http://www.un.org/Overview/rights.html (accessed 15 February 2008). For a short summary of the concept of natural law, see R.R. Palmer and Joel Colton, *A History of the Modern World*, 8th ed., (New York: McGraw-Hill, 1995), 307–13.

27 Ibid.

28 Sallie King, "A Global Ethic in the Light of Comparative Religious Ethics," in *Explorations in Global Ethics: Comparative Religious Ethics and Interreligious Dialogue*, ed. Sumner B. Twiss and Bruce Grelle (Boulder, CO: Westview Press, 2000), 136.

29 Ibid., 135–8. See also, Sumner B. Twiss, "Religion and Human Rights: A Comparative Perspective," in *Explorations in Global Ethics: Comparative Religious Ethics and Interreligious Dialogue*, ed. Sumner B. Twiss and Bruce Grelle (Boulder, CO: Westview Press, 2000), 160, 173. Twiss points out that the first article of the declaration was significantly influenced by an Asian contributor using Confucian principles.

30 Darrell J. Fasching, "From Genocide to Global Ethics by Way of Storytelling," in *Explorations in Global Ethics: Comparative Religious Ethics and Interreligious Dialogue*, ed. Sumner B. Twiss and Bruce Grelle (Boulder, CO: Westview Press, 2000), 315.

modern human rights language, should be incorporated. This is not meant to imply a final determination that the laws of man should necessarily emerge solely from a consideration of natural laws. Nor is it meant to resolve the debate with jurists who counter this tradition with that of legal positivism.[31] The intention here is merely to offer that there are elements contained within, generally agreed upon by many cultures, which offer a rational means to measure positive and negative effects as a part of the evaluative framework for this study.

Reason The seventeenth, eighteenth, and nineteen centuries, the Ages of Reason, Enlightenment, and beyond, witnessed the birth and maturation of modern Western philosophy in areas such as metaphysics, logic, ethics, and politics among others. Philosophers such as Sir Isaac Newton, René Descartes, Thomas Hobbes, and John Locke, active in the earlier period, were followed by Edmund Burke, David Hume, Immanuel Kant, Jeremy Bentham, and Voltaire in the middle.[32] Later still, John Stuart Mill, Karl Marx, Georg Hegel, and Friedrich Nietzsche continued philosophy's "long conversation."

Many of these names are quite familiar to contemporary readers, even those who are not students of philosophy, and many of the thoughts and ideas they first proposed have been internalized into modern society.[33] While seeking to fully comprehend schools of thought as varied the Rationalists, the Empiricists, and ideas such as liberalism, and existentialism, is a fascinating pursuit, it is at the same time a massive undertaking. Fortunately, two distinct ethical traditions emerge from this period that underpin much of contemporary thinking, and seem to offer value for this study. Paradoxically, these two traditions proceed from fundamental assumptions that are diametric opposites.

The ethic of duty, of which Immanuel Kant is the most famous proponent, argues that the reason an action is being done is all that really matters when making a moral decision. Alternatively, the ethics of consequences, commonly referred to as utilitarianism, first championed by Jeremy Bentham and John Stuart Mill, proceeds from the core belief that it is the outcome of our action that is of prime concern in moral decision-making.[34] Clearly neither ethic, despite what ardent supporters of either viewpoint might argue, can be unerringly correct in all situations. Merely looking at a few of the elements of the Just War Tradition, such as *just cause*, *right intentions*, *reasonable chance of success*, and *proportionality*

31 Legal positivism is the thesis that the existence and content of law depends on social facts and not on its merits. Definition from the *Stanford Encyclopedia of Philosophy*; Available at: http://plato.stanford.edu/entries/legal-positivism/ (accessed 29 February 2008).

32 Voltaire was the pen name of François-Marie Arouet.

33 Many of Locke's ideas, refinements of Thomas Hobbes earlier thoughts, were adopted as the philosophical underpinnings of the American political and judicial systems.

34 Hinman, *Pluralistic Approach*, 12–13.

demonstrates that an admixture of these two traditions is evident. Both of these traditions will now be examined in greater detail.

The Ethics of Intentions: Kant and Duty

Immanuel Kant (1724–1804) was born in Königsberg in East Prussia and, by most accounts, never traveled more than about 60 miles from that city. Although beginning his advanced education as a theology student, he became more interested in mathematics and physics, subsequently deciding on an academic career. Short of funds, he served as a tutor for nearly a decade before returning to the University of Königsberg in 1755 where, for the next 42 years, he served in a variety of academic positions of increasing seniority and responsibility. A prolific writer, albeit at times very difficult to penetrate, he authored some 19 volumes on questions such as "the mechanics of knowing; ethics; religion; and finally the nature of esthetic feeling and the direction of biological evolution. [Julien Benda argues that,] on each of these questions he expressed ideas which marked decisive turning points in the history of human thought."[35] It is his philosophy of ethics, and three key insights he provides, that is most relevant to this study.

Dismissive of the views of some other moral philosophers of the time as "a little of this and a little of that in a marvelous mixture," Kant was not concerned with outcomes.[36] He believed that, "the ordinary moral consciousness perceives quite clearly that the ethical value of an action does not depend upon its external results, but upon the interior will which gave it rise"; in other words, intentions matter, not consequences.[37] At the beginning of his foundational work on ethics, *Fundamental Principles of the Metaphysics of Morals*, Kant offered the first of his key insights, which incorporates the concept of duty into the above sentiment. He declares, "For in order that an action should be morally good, it is not enough that it *conform* to the moral law, but it must also be done for the sake of the law."[38]

In slightly more modern language, this sentiment might be expressed as "It is not enough that we do the right thing, we must do it for the right reason." This duality is critical to Kant's thinking. An individual who chooses to do the "right thing" simply because he or she knows that a personally beneficial outcome will

35 Julien Benda, *The Living Thoughts of Kant*, ed. Alfred O. Mendel, The Living Thoughts Library (Philadelphia: David McKay Co., 1940), 5–6. See also, *Kant*, Great Books of the Western World, ed. Robert Maynard Hutchins, vol. 42 (Chicago: Encyclopedia Britannica, 1952), v–vi.

36 Immanuel Kant, *Foundations of the Metaphysics of Morals: With Critical Essays*, trans. Lewis White Beck, ed. Robert Paul Wolff, The Bobbs-Merrill Text and Commentary Series (Indianapolis: Bobbs-Merrill Educational Publishing, 1969), 31.

37 Benda, *Living Thoughts*, 20; Elsewhere, "the moral worth of an action does not lie in the effect which is expected from it or in any principle of action which has to borrow its motive from this expected effect." Kant, *Foundations*, 20.

38 *Kant*, Great Books of the Western World, ed. Robert Maynard Hutchins, vol. 42 (Chicago: Encyclopedia Britannica, 1952), 254: emphasis in original.

occur is not performing a morally praiseworthy act. Kant sets a high bar here. He uses as his illustrative example the shopkeeper who is honest enough to charge all customers, even inexperienced children, the same price for his goods. While many might view this action as morally worthy, Kant argues that, "this is not enough to make us believe that the tradesman has so acted from duty and from the principles of honesty: his own advantage required it."[39] In other words, the merchant knew that if he had taken liberties with some clients, the word would have spread and his business would have, in the long run, suffered. How, then, is this laudable concept of duty to be translated into a practical guide, or set of rules, for action?

From a second key insight of Kant's, regarding the universality of moral rules, emerges an answer in the form of his famous "categorical imperative."[40] In the first section of *Fundamentals* he wrote, "I am never to act otherwise than so *that I could also will that my maxim should become a universal law*," defining a maxim as "the subjective principle of volition."[41] Later he explicitly connects maxims, universality, and the categorical imperative stating, "There is therefore but one categorical imperative, namely, this: *Act only on that maxim whereby thou canst at the same time will that should become a universal law*."[42] Hinman, using a more contemporary lexicon, explains, "A maxim for Kant is the subjective rule that a person has in mind when performing an action. The test for a maxim, then, is whether people would consistently will that all people adopt this maxim as a guide in their actions. Then it would pass the test of universalizability."[43]

Here is a case where Kant's language can cause some confusion. He seems to imply that there is but one categorical imperative, yet Kantian scholars often use the term in the plural. Rather than bog down in a philosophical discussion here, it is perhaps better to take away the main point, namely, that only maxims that pass the test of universality should be acted upon. Out of this group emerges Kant's third key insight. In a section discussing respect and using people, Hinman records Kant's reformulation of the categorical imperative on this subject. "*Act in such a way that you always treat humanity, whether in your own person or in the person of any other, never simply as a means, but always at the same time as an end*."[44]

Interestingly, although he penned the essay *Perpetual Peace* in 1795, Kant was not ignorant of the present condition of humankind in the real world. Benda notes, "Far from proving himself therein a maker of Utopias, Kant asserts that such a

39 Ibid., 258.

40 Hinman translates "categorical imperative" as "unconditional command." See Hinman, *Pluralistic Approach*, 217.

41 Kant, *Great Books*, 259–60: emphasis in original. Definition of "maxim" found in note 1. This subjective principle "must be distinguished from the *objective principle*, namely, practical law," 268 note 2.

42 Ibid., 268: emphasis in original.

43 Hinman, *Pluralistic Approach*, 218.

44 Ibid., 225: emphasis in original. Originally in Kant, *Foundations*, 54. Kant refers to this statement as a "practical imperative."

good thing as eternal peace can only very slowly come to pass and that, given what humanity still is, even the best intentioned States would show wisdom in keeping their hands on the hilts of their swords."[45] Kant's three insights examined here each offer something of value to the evaluative framework. The first, his concept of duty, doing the right thing for the right reason, and the third, never using people as merely a means to an end, provide additional philosophical weight to the just war concepts previously discussed. But, what of the second insight?

At first glance, developing a maxim specifically for the use of precision guided munitions that would pass the test of universalizability seems a difficult, if not impossible, task. Yet Kant, in *Perpetual Peace*, does provide a useful maxim. The sixth preliminary article states, "In a war one should not allow oneself to commit acts of warfare of such a kind as to make reciprocal confidence impossible when it comes time to make peace."[46] Here echoes of the concerns of Orend, and other jus post bellum thinkers, are evident, which is perhaps not surprising as he approaches the subject from a Kantian perspective! Similarly, the *jus in bello* principles of *proportionality* and *discrimination* resonate in Kant's statement. These are just the sort of enduring themes which add value and strength to the framework.

The Ethics of Consequences: Bentham, Mill, and Utilitarianism

Diametrically opposed to the evaluative criteria of the ethics of duty are the utilitarians, those philosophers for whom consequences matter, in fact "*only* consequences count."[47] To be clear, this is not the only ethical tradition that is primarily concerned with consequences. However, whereas ethical egoism focused on consequences from the perspective of the individual, shown to be problematic for this study, utilitarianism elevates the perspective to that of "everyone."[48] Hinman identifies four basic types of utilitarianism categorized by the standard of utility employed. The first utilitarian, Jeremy Bentham, coincidentally a chronological peer of Kant, championed hedonistic utilitarianism, which focused on maximizing pleasure over pain. This focus on pleasure earned Bentham his share of critics, and his philosophy "quickly came under attack as 'the pig's philosophy' because of what seemed to be its crude emphasis on sensual bodily pleasures."[49] As will be seen shortly, this reaction probably stemmed, at least in part, from a lack of a full understanding of his writings. Working a half century later his godson, John Stuart Mill, shifted the standard of utility to happiness in his ethic of eudaimonistic utilitarianism. Twentieth century philosophers again shifted the focus of utility

45 Benda, *Living Thoughts*, 28. Kant is seen as the first philosopher to explore the concept of a perpetual peace among nations, enabled by a "society of nations" perhaps similar to the United Nations.

46 Quoted in Ibid., 29.

47 Hinman, *Pluralistic Approach*, 163: emphasis in original.

48 Ibid., 63, 164.

49 Ibid., 165.

with G.E. Moore's ideal utilitarianism concerned with freedom, knowledge, and justice while Kenneth Arrow's contribution, preference utilitarianism, focused on personal preference.[50] An analysis of the first two will be sufficient to evaluate the tradition's worth for the framework.

Jeremy Bentham and hedonistic utilitarianism The son of an attorney, Bentham was sent to Oxford University to study law at what today we would consider a young age, namely 12. As with Kant, he quickly soured on his original course of study and found his own academic path. His core work on utilitarianism, *An Introduction to the Principles of Morals and Legislation*, was, by his own admission, completed by 1780 but not published until 1789.[51] Bentham opens the work by positing that "Nature has placed mankind under the governance of two sovereign masters, pain and pleasure."[52] He then carefully explains what he means by the term principle of utility:

> By the principle of utility is meant that principle which approves or disapproves of every action whatsoever, according to the tendency which it appears to have to augment or diminish the happiness of the party whose interest is in question: or, what is the same thing in other words, to promote or to oppose that happiness. I say of every action whatsoever; and therefore not only of every action of a private individual, but of every measure of government.[53]

Here, in but the second paragraph of his book, two key observations need to be made

First, while he opens the work using the term "pleasure," the word his critics will cling to, it seems clear here that he alludes to much more than mere physical sensation. In fact, in the next paragraph he expounds upon the word "utility" declaring:

> By utility is meant that property in any object, whereby it tends to produce *benefit*, *advantage*, pleasure, *good*, or *happiness*, (all this in the present case comes to the same thing) or (what comes again to the same thing) to prevent the happening of *mischief*, pain, *evil*, or *unhappiness* to the party whose interest is considered: if that party be the community in general, then the happiness of the community: if a particular individual, then the happiness of that individual.[54]

---

50 Ibid., 165–8.

51 Jeremy Bentham, "An Introduction to the Principles of Morals and Legislation," in *The Utilitarians* (Garden City, New York: Dolphin Books, 1961), 6–7.

52 Ibid., 17: emphasis in original.

53 Ibid., 17–18.

54 Ibid., 18: emphasis mine.

Bentham himself seems to have acknowledged his unfortunate choice of words, or inability to sufficiently get his meaning across to his readers later in his life. In a lengthy footnote added in July 1822 for the 1823 edition, he notes, "The word *utility* does not so clearly point to the ideas of *pleasure* and *pain* as the words *happiness* and *felicity* do."[55]

The second key observation is that here, at the very beginning of his foundational work, Bentham elevates the primary focus of utility from that of the individual to that of the corporate body, the community. While he admits that the concept of "community" is a construct, he uses the term "fictitious *body*," and that the interest of the community can only be properly viewed as "the sum of the interests of the several members who compose it," the main focus of his work is on a government's relationship to the governed.[56] He maintains that "it has been shown that the happiness of the individuals, of whom a community is composed, that is their pleasures and their security, is the end and the sole end which the legislator ought to have in view."[57]

John Stuart Mill and eudaimonistic utilitarianism A child prodigy, who "studied Greek at the age of three, Latin at eight, and wrote a history of Roman law at ten," was originally, like his father James, an adherent to Bentham's philosophy.[58] Following a bout of mental illness, he later broke from Bentham, shifting the defining quality of utilitarianism from pleasure to happiness. In his book *Utilitarianism* (1863) Mill begins by explaining why he is breaking from earlier philosophers. Likely due to the long relationships between the families, he treats Bentham gently. But, perhaps showing some of the arrogance a young and celebrated prodigy might develop, he scorns Kant's work, declaring:

> This remarkable man, whose system of thought will long remain one of the landmarks in the history of philosophical speculation, does, in the treatise in question, lay down a universal first principle as the origin and ground of moral obligation; it is this:—"So act, that the rule on which thou actest would admit of being adopted as a law by all rational beings."—*But when he begins to deduce from this precept any of the actual duties of morality, he fails, almost grotesquely*, to show that there would be any contradiction, any logical (not to say physical) impossibility, in the adoption by all rational beings of the most outrageously immoral rules of conduct. All he shows is that the *consequences* of their universal adoption would be such as no one would choose to incur.[59]

55 Ibid., 291.

56 Ibid., 18.

57 Ibid., 33.

58 *The Utilitarians* (Garden City, New York: Dolphin Books, 1961), flyleaf.

59 John Stuart Mill, "Utilitarianism," in *The Utilitarians* (Garden City, New York: Dolphin Books, 1961), 404: emphasis mine.

Having launched a proverbial "shot across the bow," he then proceeds to explain why, in general, the critics of utilitarianism have missed the point. Echoing Bentham's footnote in the later edition of his primary work, Mill argues that

> the common herd, including the herd of writers, not only in the newspapers, and periodicals, but in books of weight and pretension, are perpetually falling into this shallow mistake. Having caught up the word 'utilitarian,' while knowing nothing whatever about it but its sound, they habitually express by it the rejection, or the neglect, of pleasure in some of its forms: of beauty, of ornament, or of amusement.[60]

His point is one of context. By focusing, perhaps due to Bentham's original inability to express the fullness of the meaning he intended, on the more short term, physical manifestations or interpretations of "pleasure," readers miss the wider point the utilitarian wishes to convey. Mill himself essentially admits that utilitarian writers bear a degree of responsibility for the dilemma they find themselves in when he notes, it "must be admitted, however, that utilitarian writers in general have placed the superiority of mental over bodily pleasures ... . And on all these points utilitarians have fully proved their case; *but they might have taken the other, and, as it may be called, higher ground, with entire consistency*."[61]

Mill's unenviable task, then, was to take that higher ground, to work to counter the critics, and to reinforce the higher level focus and more permanent attributes of happiness over pleasure. While doing so he offers some additional insights. Reinforcing Bentham's concern with community, the standard, Mill observes, "is not the agent's own greatest happiness, but the greatest amount of happiness altogether."[62] Presented another way, in a more academic articulation of one of the most recognized bumper stickers of utilitarianism, he writes, "according to the 'greatest happiness principle' ... the ultimate end ... is an existence exempt as far as possible from pain, and as rich as possible in enjoyments, both in point of quantity and quality ... to the greatest extent possible, secured to all mankind."[63]

While utilitarianism seems to hold great promise to this effort, a word of caution is in order. If utilitarianism demands that we seek the greatest good for the most people, could it easily be argued that the marginalization of a given minority group would be morally justified—despite the fact that such an action might violate what we see as fair or correct? Hinman addresses such a concern using slavery as his example.[64] From a medical perspective such an approach might also be used, in the extreme, to justify, in an era of rising health care costs, the denial (or even the

60 Ibid., 406–7.

61 Ibid., 408.

62 Ibid., 412.

63 Ibid., 412–13.

64 Hinman, *Pluralistic Approach*, 184–5.

removal) of expensive medical assistance to critically ill or profoundly injured patients *even in cases where they do not wish to terminate their life.*

With that word of caution in mind, to be usefully counterbalanced by elements of Kant's philosophy perhaps, it is utilitarianism's focus on maximizing the positive aspects of people's existence, while concurrently seeking to minimize the negative aspects, however those might be defined, which demonstrates that this tradition offers much of value to the framework under development. Indeed much of American legislation is based upon just such a premise.

Character-based Systems

Up to this point, the ethical systems reviewed have principally focused on fundamental questions such as "What should I do?" or "How should I act?" There is, however, another approach which seeks answers to questions such as "How should I live?" and "What kind of person should I be?" Because this study is using a thematic approach to these ethical systems, as opposed to a historical linear approach, it now turns to a much earlier writer, the Greek philosopher Aristotle (384–322 BCE) who first brought these ideas to prominence.[65]

Aristotle was born in the city of Stagira in Chalcidice, the triple peninsula in Macedonia, into the family of Nicomachus, court physician to Amyntas II. His parents died while he was young and at the age of 17 he entered Plato's Academy, where he stayed for the next two decades, leaving upon Plato's death. Later tutor to Alexander the Great, he eventually returned to Athens in 335 BCE where he opened his own school, the Lyceum (later called the Peripatetic School), where he remained until shortly before his death. Although he gave lectures, he spent much of his time writing and tutoring small groups of students, eventually leaving a large body of work to the world.[66] While noted for his many writings on logic, metaphysics, biology, politics, and rhetoric among others, it is the thoughts contained in his two works on ethics, *Nicomachean Ethics* (*Ethica Nicomachea*) and *Eudemian Ethics* (*Ethica Eudemia)*, that are of specific interest to this study.[67]

Aristotle and the concept of flourishing In *Aristotle on Ethics*, Gerard Hughes offers an explanation as to why Aristotle wrote *Nicomachean Ethics*. He believes that Aristotle wrote this treatise for the same reason that Plato wrote his *Republic*,

65 In the West, our ethical heritage is often seen as descending from ancient Greece. This focus on Aristotle is not intended to diminish or ignore the contributions of other Greek philosophers such as the Stoics. While not explicitly addressed earlier, stoicism gave rise to the philosophy of natural law/natural rights which eventually influenced Thomas Jefferson as he wrote the Declaration of Independence.

66 Biographical details drawn from *Oxford Classical Dictionary*, 3rd ed. (1999), s.v. "Aristotle."; See also *Aristotle: I*, Great Books of the Western World, ed. Robert Maynard Hutchins, vol. 8 (Chicago: Encyclopedia Britannica, 1952), v–vi.

67 There is significant overlap of material between these two works.

specifically, that "He hoped to provide an account of how the good person should live, and how society should be structured in order to make such lives possible," an assertion which seems to be supported by a passage early in Book I.[68] There Aristotle declares, "though it is worthwhile to attain the end merely for one man, it is finer and more godlike to attain it for a nation or for city-states. These, then, are the ends at which our inquiry aims."[69] At this point, Aristotle is just beginning to set up the logic of his argument and uses the term "end," however, as it soon becomes clear, he is referring to the "good life" or "the life worth living."

If this is accurate, then it seems worthwhile to examine what constitutes a "good life." Why is that important to this particular effort? The answer rests upon one fundamental realization. If we can understand what Aristotle means by a "good life," and if we accept his position, then we can evaluate political and military actions, planned outcomes, and/or unintended consequences as either advancing (a positive trend) or retarding (a negative trend) the "good life," thereby adding another component, from a different philosophical viewpoint, to the evaluative framework.

For Aristotle, the answer to what constitutes the good life is deceptively simple, although it takes most of the volume for him to work through all the details, nuances, and refute possible counter-arguments. Simply stated,

> Let us resume our inquiry and state, in view of the fact that all knowledge and every pursuit aims at some good, what is it that we say political science aims at and what is the highest of all goods achievable by action. Verbally there is very general agreement; for both the general run of men and people of superior refinement say that *it is happiness, and identify living well and doing well with being happy.*[70]

At first glance it would appear that Aristotle might have fallen into the same linguistic morass that Bentham was to fall into so many centuries later. But he knows the difficulties this word might present, noting that "For [some] think it is some plain and obvious thing, like pleasure, wealth, or honour."[71] If this is the case, then how should we construe the word "happiness" in this context?"

To begin with, the Greek word used in the original, and most commonly translated as "happiness," is *eudaimonia* (εὐδαιμονία).[72] However, a number of authors have provided a nuanced interpretation arguing, for example, that,

68 Gerald J. Hughes, *Routledge Philosophy Guidebook to Aristotle on Ethics*, Routledge Philosophy Guidebooks, ed. Tim and Jonathan Wolff Crane (London: Routledge, 2001), 13.

69 Arist., *Eth. Nic.*, I, 2, 1094b8-10.

70 Ibid., I, 4, 1095b13-19.

71 Ibid., I, 4, 1095b22-23.

72 Defined as "happiness or well-being consisting in the full realization of human potential." *Oxford Dictionary of Foreign Words and Phrases*, (1997), s.v. "eudaimonia."

"Aristotle's ethical treatises search for an adequate account of *eudaimonia*, a term usually translated 'happiness,' but which might more perspicuously [i.e. more clearly or insightfully] be understood as 'human flourishing.'"[73] With a number of definitions of the words flourish and flourishing containing words such as "thrive," "prosper," "do well," and "to be in one's bloom or prime," the *OED* can be seen as adding support to this nuanced view.[74] In fact, later in *Nicomachean Ethics*, Aristotle himself appears to embrace this interpretation when he states, "If happiness is activity in accordance with virtue, it is reasonable that it should be in accordance with the highest virtue; and *this will be that of the best thing in us*."[75]

There is much more to Aristotle's philosophy that is valuable to us, such as his discussions of virtue and of leading a balanced, temperate life, but for this study the key insight for the framework has been obtained. The correct life to be lived should enable individuals, and the community as a whole, to realize their full potential. What this adds to this study is an accepted philosophical concept that can be used, in conjunction with insights gained from natural law, utilitarianism, and other philosophies, to help determine whether given events should be viewed in a positive or negative light based on how they affect human flourishing.

This selective examination of ethical traditions leads us to the next step, which is to synthesize the insights gained into an effective and workable framework that can be used to evaluate the case studies selected. As this next step is taken, there is a key point worth making. This overall effort to bring ethical theory to bear on a particular modern technological application, PGMs, will require that the framework blend technical detail with ethical theory, a necessity that may seem awkward at first.

The Evaluative Framework

As previously discussed, the analytical tool under development is divided into three parts. The first looks at the weapons and their employment, while the second examines the issue of unintended consequences, and the third explores how having this capability has impacted the national leadership and its decision-making process relative to the use of military force. While each will be examined in isolation, it must be recognized that no such clear delineation exists in the real world. The dividing line between each part of the framework is blurry at best, and a given factor, such as targeting strategy, may impact the analysis in more than one area. But as with many academic endeavors, such as mechanical engineers studying statics before they study dynamics, dividing complex tasks into parts helps to enable greater understanding of the whole once reconstituted.

73 *Oxford Classical Dictionary*, 3rd ed. (1999), s.v. "Aristotle."
74 *Oxford English Dictionary*, 2nd ed. (1989), s.v. "flourish/flourishing"
75 Arist., *Eth. Nic.*, X, 7, 1177a11-12: emphasis mine.

Before proceeding, it is worth mentioning why developing an effective framework with which to measure the impact that PGMs have had on human suffering in war is a difficult proposition. First, while we are examining the impact of a specific technology, PGMs in this case, on the human condition, it must also be acknowledged that this technology is but one prime enabler of what many see as a new style of warfare. As a result, care must be taken to discriminate, wherever possible, between the impact this particular technology is having on human suffering and the impact other technologies, also enablers of this new manner of warfare, may be having.[76] Additionally, the framework is being designed with opportunities for future study in mind. If carefully crafted, minor modifications to the lines of inquiry and metrics to be used should allow the framework to be used to evaluate other emerging technologies.

Finally, building an effective framework is difficult because it is not inherently a quantitative study. The solution is not as simple as developing a number of numerically based metrics which can be easily measured and placed into a clever algorithm to provide a final, definitive answer. True, numbers will provide insight at times, but the study, at its core, seeks to identify trends, and specifically, trends caused by a given technology, as well as how that technology has influenced the senior leaders who employ it to advance national objectives.

Part I: The Weapons and Their Employment

Two of the key principles within just war are proportionality and discrimination. One of the long standing difficulties with aerial bombing that directly impact these two principles, and the one that has caused the most human suffering, is that of accuracy. As previously explained, accuracies for munitions are usually expressed in terms of a Circular Error Probable or CEP. By definition, this means that 50 percent of weapons released will fall within that stated distance. Many factors come into play in calculating this number including weapon design, delivery parameters, aircrew proficiency, and the simple fact of whether the aircrew is under threat from enemy defenses. In truth, CEPs reported during peacetime training, particularly for non-guided weapons, are much lower than those actually realized in combat.[77]

At one point during World War II, for example, "the British measured night-bombing accuracy in miles while the Americans measured their daylight, visual

76 An illustration of another technology might be "cyber warfare." For example, during air strikes, PGMs might take out a number of key infrastructures while a computer virus destroyed the banking system of a nation. Discriminating between the two, if possible, would be important to keeping the evaluation focused.

77 For an excellent analysis, see Lt Col Bernard Appel, USAF, "Bombing Accuracy in a Combat Environment," *Air University Review* July/August 1975.

accuracy in thousands of feet."[78] Two decades later, the CEP for non-precision weapons in Vietnam was calculated as 447 feet.[79] Sidestepping the debate of the moral correctness of aerial bombing when accuracies are known to be so poor, it is important to note that such figures highlight a number of harsh realities related to bombing.

First, such large inaccuracies invariably mean numerous off-target impacts resulting in collateral damage to the surrounding area, including both physical structures and people. Next, to counteract the low probability of destroying the desired target with any given aircraft, planners are forced to "throw" more airplanes at the problem thus increasing the number of aircrew at risk from enemy defenses while, concurrently, creating even more off-target impacts. For example, the British Bomber Command losses alone during the strategic bombing campaign in Europe during World War II stand at some "8,325 airplanes destroyed and almost 64,000 aircrew casualties."[80] Finally, as was seen in World War II, although certainly less since then, planners sometimes chose as a policy the direct targeting of civilian populations in order to influence the outcome of the conflict.[81]

Two significant benefits that may be realized with the employment of precision-guided munitions are that of superior accuracy, and greater target discrimination. Anecdotal evidence in the last decade or so would seem to bear out these promises. Aircrews, and weapons manufacturers, often talk of "targets per sortie" instead of the old measure of "sorties per target." Publicly released CEPs for the latest generation of PGM, the Joint Direct Attack Munition, or JDAM, are as low as 5 to 12 meters. Additionally, with the greater stand-off distance achievable with these newer weapons, aircrew survivability has improved.[82]

But if the remainder of this study is to have credibility, it must first validate these anecdotal claims. Simply put, do PGMs work as advertised? Therefore, the first task is to examine the performance and employment of the weapons themselves. Here two distinct, but deeply interrelated, questions need to be answered. First, "Have these weapons been as operationally effective and discriminate as has been advertised, thereby reducing human suffering?" and, second, "Have these

---

78 Kenneth P. Werrell, "The Strategic Bombing of Germany in World War II: Costs and Accomplishments," *The Journal of American History* 73.3 (December 1986): 706.

79 Max Boot, "From Saigon to Desert Storm: How the U.S. Military Reinvented Itself After Vietnam," *American Heritage Magazine* 57.6 November/December 2006.

80 Werrell, "Strategic Bombing," 708.

81 Melden E. Smith, Jr., "The Strategic Bombing Debate: The Second World War and Vietnam," *Journal of Contemporary History* 12.1 (January 1977): 183. Smith also highlights why we need to move past "popular mythologies" when studying matters of import. Although widely accepted that some 135,000 people perished in the Dresden firebombing, East Germany officially set that number at 35,000. Not trivial, but significantly lower than popularly believed.

82 The USAF JDAM fact sheet can be found at: http://www.af.mil/factsheets/factsheet.asp?fsID=108.

weapons actually been employed in a manner specifically designed to reduce human suffering to the greatest extent possible?"

Operational performance and effectiveness The first portion of the analysis may prove to be the easiest of the entire study, and the one most dependent on a quantifiable measure. Given the level of post-strike/conflict analysis now routinely conducted by the Department of Defense, the actual performance of PGMs should be able to be accurately determined by their demonstrated CEP and by an analysis of the weapons failure rate. If actually observed combat CEPs are indeed as low as advertised, and the number of off-target impacts less than previously experienced as a result, then PGMs would, for this part of the analysis, be seen in a positive light in terms of lessening human suffering. Higher accuracy and lower levels of collateral damage would satisfy *jus in bello* concerns with proportionality and discrimination.

Manner of employment Closely related to the delivery accuracy of the weapons is the planned manner of their employment. Here issues of targeting strategy are preeminent. Are the weapons scheduled to be employed at such a time or in such a manner so as to minimize the loss of life of non-combatants? While the principle of double-effect may make such deaths morally acceptable, albeit regrettable, it cannot be denied that some sense of guilt is attached to each of these instances. One example of humane planning would be the targeting of a key building after normal business hours when most of the work force has departed.[83]

Another issue is that of "no-strike" lists. Governments routinely issue directives to planners preventing certain targets from being attacked. For a number of reasons, hospitals, schools, religious sites, and sites of great cultural significance are exempted.[84] If such planning measures are in place, and adhered to, then such measures would be viewed positively in regards to human suffering. Here, again, the *jus in bello* criteria of discrimination helps underpin this determination. But elements of utilitarianism and natural rights are also present, as careful targeting will ensure that the lowest amount of suffering as possible is inflicted on the populace, who remain free to go about their lives.

The measured employment of PGMs also directly addresses one of the key concerns that Ellul has with modern technology. It is clear that the drive to greater weapons accuracy is a clear example of Ellul's concern that "technical progress today is no longer conditioned by anything other than its own calculus of efficiency."[85] But the active participation of the human element in the planning process, seeking to temper the unthinking application of PGMs in war, would seem

83 Attacking at night decreases the likelihood that the delivering platform will be detected engaged visually.

84 At times to the detriment of U.S. forces as the enemy may chose to engage from such sites knowing that U.S. military forces are not likely to strike them.

85 Ellul, *Technological Society*, 74.

to represent an example of that individual who "can doubtless seek the soundest attitude to dominate the techniques at his disposal."[86] A version of Postman's "loving resistance fighter" perhaps?

Table 3.1 summarizes this line of inquiry. While answers to the questions in this part of this framework will provide considerable insight into resolving the study's overarching question, they only address the issue of human suffering in the short term, primarily during the period of actual combat operations.

Part II: The Question of Unintended Consequences

Unfortunately, the decision to employ precision weapons to seek a decisive outcome with lower levels of violence may also have broader ethical consequences in the longer term. Therefore, to be comprehensive, the analysis must deliberately embrace an expanded timeframe. What is examined in this second part of the framework is the issue of second order effects, or unintended consequences, those additional outcomes which were unanticipated by those who used PGMs, and which may fundamentally alter the conclusions logically drawn from the first part of the analysis.

Table 3.1 Weapons effectiveness and operational employment

Line of Inquiry	Philosophical Underpinnings	Metrics	Trend Line
1. Have these weapons been as operationally effective and discriminate as has been advertised, thereby reducing human suffering?	Just War: *jus in bello*	Circular Error Probable (CEP)	Lower CEP: Positive Higher CEP: Negative
		Weapons Functional Effectiveness	Higher %: Positive Lower %: Negative
2. Have these weapons actually been employed in a manner specifically designed to reduce human suffering to the greatest extent possible?	Just War: *jus in bello* Utilitarianism	Evidence of deliberate planning to reduce casualties and collateral damage?	Yes: Positive No: Negative
	Natural Rights *Technique*	Evidence of appropriate "no strike" considerations?	Yes: Positive No: Negative

86 Ibid., 21–2.

Such a reassessment may be brought about by a number of situations, three of which are as follows. First, does the use of such weapons result in a more rapid capitulation of the enemy leaving a greater portion of underlying societal and/or political problems unresolved? Approaching from a slightly different perspective, the noted naval analyst Norman Friedman wondered if using such precise means, left adversaries "undefeated" psychologically, leading to more resistance and greater human suffering in the long run.[87] France during World War II, post-June 1940, might provide an historical example of such a situation, albeit as a result of a different "new way of warfare." In roughly six weeks, using its innovative brand of mobile warfare, Germany defeated the armies of France, which surrendered on 22 June 1940.[88] Compared to the interminable trench warfare of World War I, this period could be seen as having a relatively low level of human suffering. Yet the rapid defeat left the French populace unsubdued, leading to the rise of the resistance movement in German-occupied Vichy France. Attacks on Germans, and the resultant reprisals against the French, continued until the country's liberation some four years later.

Next, does reliance on such highly technical solutions force our adversaries to adopt lower tech, asymmetric responses which, paradoxically, increase levels of human suffering compared to more conventional force on force engagements? Unable to confront the U.S. military directly, suicide bombers immediately spring to mind when pondering this question. So too, perhaps, do the attacks on American embassies around the world, the USS Cole, and the terrorist attacks of 9-11.

Finally, does the use of such weapons, while reducing initial levels of human casualties, actually result in greater suffering in the long term due to damaged critical infrastructure, both physical and organizational? A government, for example, might choose to use PGMs, at night, to strike the generators at a local power station located in a densely populated part of a city to shut down military communications networks and air defense radars. Such an action would likely kill or injure few at the power plant or the surrounding neighborhoods, particularly when compared with the former need to use massed "dumb bombs" to complete the task, and at the same time, would likely result in lower attrition for the attacking aircraft. This would be a relatively positive event according to the framework as there would be little human suffering involved.

However, an unintended consequence of that action might be the simultaneous elimination of power to critical life support equipment at regional

87 Norman Friedman, "Is Modern War Too Precise?" *Proceedings* 130.12 (December 2004): 4. Clausewitz stated the results of war are never final, that "the defeated state often considers the outcome merely as a transitory evil, for which a remedy may be found … at some later date" (von Clausewitz 1976, 80). Friedman uses the case of Germany following the two World Wars as examples.

88 The term "blitzkrieg" was first popularized in a September 1939 article in *Time* magazine.

hospitals and nursing homes, structures normally exempt from attack. If repairs could not be affected quickly, or back up power was not available, the initial "surgical" action eventually would result in a significant amount of suffering in the civilian population forcing an ethical reassessment of the action. This is but one hypothetical example where the use of precision-guided munitions, to use Tenner's term, "bites back." This also represents one of Jacques Ellul's key concerns. Simply put, he is worried that in an era when technological *means* become the focus, to the relative exclusion of rationally considering *ends*, "certain unforeseen secondary effects ... are much more disastrous than the lack of technique would have been."[89] An important part of this inquiry is to determine if he is correct.

But how best to identify those metrics to be used to evaluate these second order effects and unintended consequences? Bentham and Mill, in their long explications on utility, identify a number of sources of pleasure and happiness. Bentham, for example, identifies four, the physical, the political, the moral, and the religious.[90] So too, Aristotle offers insight as to areas where humans flourish. Similarly, a number of documents, such as the NATO Charter and the Declaration of Independence, address things like "life, liberty, and the pursuit of happiness." Which standards are most acceptable to the most people and, therefore, should be adopted? A satisfactory answer may be found in the UN's *Universal Declaration of Human Rights*, previously discussed. While certainly not a perfect solution, since clearly not every culture or society accepts every article without question or qualification, as a consensus document from a world body using natural/human rights contemporary language it has much to offer to this pluralistic effort. This language is affirmative, speaking repeatedly of the right of "all members of the human family" to be free of any attempt to prevent them from leading fulfilled lives, the ability to flourish. Several articles contain language for which trends can be readily observed and thus can be usefully adopted as standards for the framework.

For example, *Article 3* states, "Everyone has the right to life, liberty and security of person." *Article 5* contains a prohibition against "torture ... cruel, inhuman or degrading treatment or punishment." *Articles 7* through *12* outline equal protections for all under the rule of law. The first part of *Article 13* outlines a "right to freedom of movement and residence within the borders of each state." *Article 17* declares "No one shall be arbitrarily deprived of his property." *Article 23* outlines the belief that "everyone has the right to work ... [and] to just and favourable conditions of work and to protection against unemployment." Finally, *Article 25*, in particular, provides a number of useful metrics as it states,

> Everyone has the right to a standard of living adequate for the health and well-being of himself and of his family, including food, clothing, housing and

89 Ibid., 105.

90 Bentham, *Principles*, 33.

> medical care and necessary social services, and the right to security in the event of unemployment, sickness, disability, widowhood, old age or other lack of livelihood in circumstances beyond his control.[91]

Table 3.2 summarizes the analytic effort for this portion of the framework. For each case study, those actions or events that enable humans to flourish across the range of standards selected, resulting in less human suffering are viewed as positive trends. In a similar fashion, those actions that inhibit or reduce flourishing, increase human suffering and will be viewed negatively, although of course not every standard will be relevant in each case study.

Part III: The Question of Teleology

Having examined the operational effectiveness and employment of the weapons themselves, and the impact of any unintended consequences, there is one final issue that must be explored as it, too, may alter the conclusions logically drawn from the first two. Here the analysis seeks to understand the impact the availability of precision-guided munitions technology has on the national leadership and its decision-making process relative to the use of military force, a teleological question. Simply put, how does the use of PGMs, which are a means, influence the consideration of ends? Again, there are two lines of inquiry to follow in the effort to find as complete an answer as possible.

First, does the relative availability of such weapons, and their apparent effectiveness, breed reluctance in American leaders to employ historically more casualty-intensive conventional military forces, thereby opening up opportunities for our adversaries? This is one question that may best be illuminated by the Balkans case study.

It will be interesting to determine if the public announcement by the U.S. government that ground troops would not be introduced into the theater of operations opened up a window of opportunity for the adversary to pursue a course of action that increased human suffering in the region.

The second question, perhaps the more critical of the two, ponders whether, because senior military and civilian leaders have confidence in the use of such weapons, we use lethal force more often than in the past? In other words, "Does possessing the ability to strike an enemy with little collateral damage encourage leaders to embark on military operations, outside the classic definition of war but within the Clausewitzian understanding, before *jus ad bellum* criteria are satisfied?" Such a two-pronged approach engages the ethics of Bentham, Mill, and Kant, for while it is true that a utilitarian approach seeks to ensure the greatest good for the greatest number of people, we must also remember our duty to allow the least among us the opportunity to flourish.

---

91 Universal Declaration of Human Rights (1948). Available at: http://www.un.org/Overview/rights.html.

Table 3.2 Second order effects (unintended consequences)

Line of Inquiry	Philosophical Underpinnings	Metrics	Trend Line
1. Does the use of such weapons result in a more rapid capitulation of the enemy leaving a greater portion of underlying societal and/or political problems unresolved?	Just War: jus post bellum Utilitarianism Technique	Evidence of resistance type activity after major combat over *or* significant unresolved inter-/intra-state political problems	Yes: Negative No: Positive
2. Does our reliance on such highly technological solutions force our adversaries to adopt lower tech, asymmetric responses which, paradoxically, increase levels of human suffering?	Utilitarianism *Technique*	Evidence of violent asymmetric response by adversary	Yes: Negative No: Positive
3. Does the use of such weapons, while reducing initial levels of human casualties, actually result in greater suffering in the long term due to damaged critical infrastructure, both physical and organizational?	Just War: jus post bellum Natural Law/Rights Virtue (Flourishing) Utilitarianism Technique	U.N. UDHR—The right to: *Art. 3* (life, liberty) *Art. 5* (no inhuman treat.) *Art. 7-12* (rule of law) *Art. 13* (free movement) *Art. 17* (property) *Art. 23* (work) *Art. 25* (quality of life)	For each of these seven areas, outcomes which bring people closer to the rights articulated in a given article are *positive.* Those which deny people those said rights are seen as *negative.*

This third part of the framework may be the most difficult to draw meaningful conclusions from, although the effort must be made to ensure thoroughness. Necessarily, it will involve some conjecture. For example, in the past, before this capability existed, a president may have shied away from using military force in a given crisis situation due to concerns about the cost in "blood and treasure." The fact that he or she could, and would, employ force today seems, at first glance, to indicate a negative trend as the suffering caused by that action would otherwise not have occurred. But the possibility must also be considered that the use of force today, with its attendant level of suffering, might actually prevent or forestall greater suffering in the future. If this were the case, then the action would be seen as positive. A difficult conclusion to reach, but it must be wondered if some small military action early on in Rwanda, for example, could have helped the world avoid the genocide there. Table 3.3 summarizes this part of the framework.

Table 3.3 The question of teleology: A consideration of ends

Line of Inquiry	Philosophical Underpinnings	Metrics	Trend Line
1. Does the relative availability of such weapons, and their apparent effectiveness, breed reluctance in American leaders to employ historically more casualty-intensive conventional military forces, thereby opening up opportunities for our adversaries?	Duty Utilitarianism	Evidence of greater suffering due to a reluctance to deploy forces	Yes: Negative No: Positive
2. Does possessing the ability to strike an enemy with little collateral damage encourage leaders to embark on military operations, outside the classic definition of war but within the Clausewitzian understanding, before *jus ad bellum* criteria are satisfied?	Just War: *jus ad bellum* Duty Utility Virtue	Evidence of a greater willingness to use force due to PGMs advanced capabilities	Yes: Negative, unless it can be persuasively demonstrated that greater future suffering was alleviated by the action No: Positive

While certainly not perfect in every application, for over two hundred years America has sought to maintain a moral high ground in its national and international dealings. Such sentiments as "we hold these truths to be self evident, that all men are created equal, that they are endowed by their Creator with certain unalienable Rights, that among these are Life, Liberty and the pursuit of Happiness" are embodied in its founding documents and seem to be a more modern articulation of the Greek concept of *eudaimonia* or "flourishing." If the nation truly believe this, then it seems essential that it test itself against these high ideals. In the final analysis, this study, examining the most deadly of human endeavors, seeks to determine whether America is, in fact, enabling these rights to the greatest extent possible in terms of the weapons it develops and the manner in which it employs them. In this complex world full of competing priorities, is the nation maintaining a balance between the need to employ lethal force and enabling the fulfillment of these "unalienable rights?" Analysis of the following case studies will help determine if this is the case.

Chapter 4

Vietnam: Viable Precision Emerges

> The unparalleled, lavish use of firepower as a substitute for manpower is an outstanding characteristic of U.S. military tactics in the Vietnam War.
>
> Robert M. Kipp, "Counterinsurgency From 30,000 Feet: The B-52 in Vietnam"

Coming just two decades after the massive carnage of the Second World War, and a little more than a decade after the first proxy conflict of the new Cold War era, the war in Southeast Asia sorely tested the cohesion of the Clausewitzian trinity for the United States.[1] Over time, the military, increasingly frustrated by its inability to successfully find and decisively defeat the enemy, chafed under limitations set by politicians in Washington, limitations that, however frustrating, were fully in keeping with our system of political primacy in state affairs.[2] The American public, initially supportive of the war, came to believe its government was lying, a distrust seemingly validated by later revelations and leading to massive anti-war demonstrations at home.[3] Returning soldiers, many already scarred by combat operations in theater, were, in many instances, shunned by the wider civilian population as "baby killers," left to wonder how serving their country, as generations before them had when called, made them "bad."

To be fair, it was always complicated war, one that noted Vietnam historian, George Herring, believed "may have been the most complex war ever fought by the United States."[4] It was a curious admixture of a local insurgency, a civil war between two parts of a divided Vietnam, a conventional struggle of wills between a

---

1 The three sides of the Clausewitzian "paradoxical trinity" are "composed of primordial violence, hatred, and enmity, which are to be regarded as a blind natural force; of the play of chance and probability within which the creative spirit is free to roam; and of its element of subordination, as an instrument of policy which makes it subject to reason alone." He goes on to say, "The first of these aspects *mainly* concerns the people; the second the commander and his army [contemporarily extended to include the entire military]; the third the government." Clausewitz, *On War*, 89: emphasis added.

2 Many have written about these civil-military tensions. For one insight, see George C. Herring, *LBJ and Vietnam: A Different Kind of War*, An Administrative History of the Johnson Presidency Series (Austin: University of Texas Press, 1994), 51–7.

3 In February 1965 18 percent of those polled thought the U.S. should pull its forces out of South Vietnam. But, McNamara later noted, "these numbers changed dramatically over the next three years, as Johnson's continued lack of candor steadily diminished popular faith in his credibility and leadership." Robert S. McNamara with Brian VanDeMark, *In Retrospect: The Tragedy and Lessons of Vietnam*, (New York: Time Books, 1995), 173.

4 George C. Herring, "In Cold Blood: LBJ's Conduct of Limited War in Vietnam." *The Harmon Lecture Series* in Military History #33, United States Air Force Academy, 17

large industrialized nation and a much smaller, more agrarian society, all of which was waged against the backdrop of the continuing ideological struggle between East and West and a burgeoning civil rights movement back home in the United States.[5] Yet this ill-fated war, with its horrors and failures seared into the collective psyches of a generation of servicemen and civilians alike, serves as a bridging case study between two eras of warfare, and also serves as the progenitor of the style of warfare so prominently displayed in modern American combat operations.

Initially fought with the basic weapons and tactics of the two preceding wars, more advanced technologies began to emerge. The guns of Korean-era fighter aircraft were gradually augmented with longer-ranged air-to-air missiles. Similarly, anti-aircraft artillery (AAA) was enhanced by the introduction of early radar-guided surface-to-air missiles (SAMs). Most importantly for this study, the age of aerial bombardment employing unguided, "dumb" bombs began a slow transformation, to be replaced by the age of operationally effective precision-guided munitions.

With first operational employment in 1967, these more modern PGMs, in a variety of forms, assumed an ever-increasing and important role in the bombing campaigns of Southeast Asia. These weapons seemed finally to deliver the promise of early air power advocates, namely the ability to influence military operations from the air and thus, potentially, reducing the need for massed ground forces. This shift eventually became part of what some writers later termed a Revolution in Military Affairs (RMA), a trend that influences the nation's senior political and military leadership and that continues to the present day.[6]

The Air War in Southeast Asia

The opening foray in the air war in Southeast Asia did not begin, as is often believed, with the reprisal strikes launched from the U.S. Navy aircraft carriers *Ticonderoga* and *Constellation* following North Vietnamese torpedo boat attacks in the Gulf of Tonkin on 2 and 4 August 1964. American advisors had already been flying combat sorties, under the guise of "training missions," in South Vietnam. In fact, between 1962 and 1964 more than 14,000 sorties were flown with some 86 aircraft lost to

October 1990, 2. History will decide if the more recent conflicts in Iraq and Afghanistan will capture that descriptor.

5 McNamara observed that, "One reason the Kennedy and Johnson administrations failed to take an orderly, rational approach to the basic questions underlying Vietnam was the staggering variety and complexity of other issues we faced." McNamara, *Retrospect*, xvii.

6 Lacquement notes that "The RMA is characterized by a variety of high-technology features, such as the increased use of information, robotics, *precision-guided weapons*, and precise, real-time reconnaissance *tied to attack platforms that place few if any lives at risk*. Lacquement, *Shaping*, 23: emphasis added.

hostile action.[7] Additionally, American air reconnaissance operations into portions of Laos began as early as the spring of 1964.[8] But this initial offensive action into North Vietnam was to have tremendous strategic consequences for both the American military and the nation as a whole.

Taking place in the early morning hours of 5 August 1964, these strikes targeted the four home bases of the torpedo boats and associated fuel storage facilities.[9] Two days later, the joint "Tonkin Gulf Resolution," overwhelmingly passed by both the House and the Senate, granted the president the authority "as Commander in Chief, to take all necessary measures to repel any armed attack against the forces of the United States and to prevent further aggression."[10] Despite this authority, for the next seven months, with just two exceptions, no additional offensive operations were authorized into North Vietnam.

Instead, a spirited debate raged within the Johnson administration, coincidental to the 1964 presidential campaign and beyond, over the proper course of American policy in Southeast Asia, and the role that air power would or could play in it. During this period, various alternatives ranging from intensive U.S. air operations into North Vietnam, to the complete withdrawal of U.S. forces from the South, were discussed between key advisors. While most agreed that the political situation in South Vietnam was tenuous at best, consensus was not to be found.

Especially noteworthy were the fundamental disagreements as to the effectiveness of air power, particularly in this geo-political context. Those who were convinced of the efficacy of air operations against the North strongly held that such operations would destroy "Hanoi's will to fight and its ability to continue to supply the Vietcong."[11] In support of such a potential air campaign, a list of 94 targets was drawn up by the Joint Chiefs of Staff.[12] Others, however, were decidedly less convinced.

General Harold K. Jones, then-Army Chief of Staff, "argued that the rationale for air strikes was gravely flawed."[13] His viewpoint seems supported by the results

7 Guenter Lewy, *America in Vietnam* (New York: Oxford University Press, 1978), 24, Table 1-1.

8 Raphael Littauer and Norman Uphoff, eds., *The Air War in Indochina*, Revised (Boston: Beacon Press, 1972), 76. Operations in Cambodia would not begin until 1970.

9 Lewy, *America in Vietnam*, 33. For this study, it is immaterial whether the attacks of 2 and 4 August 1964 transpired exactly as then described. While historians generally agree that the 2 August attack against the *Maddox* occurred, less certainty surrounds the latter. However, as Lewy clearly outlines, *at that time* decision-makers believed American forces had been attacked twice; McNamara, *Retrospect*, 128, 135.

10 Ibid. The resolution was approved in the Senate by a vote of 88–2 and by a unanimous 416-0 in the House. It was repealed in January 1971; McNamara, *Retrospect*, 138.

11 McNamara, *Retrospect*, 152; Lewy, *America in Vietnam*, 41, 374; Pape, *Bombing to Win*, 175.

12 Ibid. Airfields, lines of communication (LOCs), military and industrial organizations, and armed reconnaissance routes were on this list.

13 McNamara, *Retrospect*, 152.

of the Joint Staff's war game "Sigma II-64," conducted mid-September 1964, which "concluded that 'industrial and military bombing' of North Vietnam 'would not quickly cause cessation of the insurgency in South Vietnam' and, indeed, 'might have but minimal effect on the (low) living standard' of the adversary."[14] In an April 1965 cable to Secretary of State Dean Rusk, Ambassador Maxwell Taylor argued, "No amount of bombardment ... is going to convince Hanoi to call off its action ... without real progress in South Vietnam against the VC. Hanoi must be convinced that the VC cannot win here."[15] As then-Secretary of Defense Robert S. McNamara eventually confessed, "This question of bombing effectiveness ... became a fundamental issue between the president and me, on the one hand, and the chiefs and military commanders in Vietnam on the other, for the next three and a half years."[16]

Just before Election Day in November 1964, a high-level working group under the leadership of Assistant Secretary of Defense for International Security Affairs William P. Bundy was set up by the president to review, develop, refine, and analyze the policy alternatives. A month later, the group presented Johnson with three options:

> A. Continuing the present course indefinitely with little hope of avoiding defeat.
> B. Undertaking a sharp, intensive bombing campaign against North Vietnam's communications lines to the South and the 94 targets proposed by the chiefs, with the object of forcing Hanoi to stop supporting the Vietcong and/or enter negotiations.
> C. Undertaking the same bombing campaign in a graduated manner, with the same objectives but a lesser risk of a larger war.[17]

Since at least the fall of 1964, the proposed air operations into North Vietnam were envisioned by most as a graduated response option, as reflected in Option C above. In November, for example, the Joint Chiefs of Staff renewed "its recommendation for strikes against the North ... in terms of 'a controlled program of systematically increased military pressures.'"[18] In Principals meetings towards the end of the

14 Ibid., 153.

15 Ibid., 182.

16 Ibid.

17 Ibid., 159–62; The option of simply walking away from the situation in South Vietnam was no longer on the table, as it was seen as being in conflict with the commitment of National Security Action Memorandum 288 (17 March 1964) to an independent, non-communist South Vietnam. U.S. Department of Defense, *United States—Vietnam Relations 1945–1967*, vol. 4, (Washington, DC: U.S. Government Printing Office, 1971), sec. IV.C.2.(c), i. Hereafter referred to as the "*Pentagon Papers.*"

18 Quoted in Department of Defense, *Pentagon Papers*, vol. 4, sec. IV.C.2.(c), ix. The source document is Joint Chiefs of Staff Memorandum 967-64 dated 18 Nov 1964.

month, the State Department and the Office of the Secretary of Defense also supported Option C, as did the U.S. Ambassador to South Vietnam, Maxwell D. Taylor, who was back in the United States for consultations.[19]

By early spring, and despite his misgivings, the president felt compelled to act, driven largely by the recognized need to boost the morale of a failing allied government in South Vietnam.[20] He ordered the commencement of what quickly became sustained air operations against the North. This action, coupled with the introduction of American ground combat forces less than a week later, began America's expanded, and extended, engagement in Southeast Asia.[21]

The sheer level of effort of the extended air war in Southeast Asia is astonishing. One study, using data provided by the Department of Defense, calculated that roughly 6,300,000 tons of aerial munitions (bombs etc.) were dropped from the beginning of sustained operations in 1965 until the end of 1971, more than three times the total dropped by the U.S. in World War II and six times the total dropped during the Korean conflict.[22] At various times during the war, air operations took place in North and South Vietnam, as well as into adjoining Cambodia and Laos. Operations in each of these geographic areas offer useful insights for the study's analysis. The operations into North Vietnam will be examined first.

Aerial Operations in the North

The aerial bombing campaign into North Vietnam can be readily divided into three distinct periods between 1965 and 1972. Three code-named operations, ROLLING THUNDER (2 March 1965–31 October 1968), LINEBACKER I (10 May 1972–23 October 1972), and LINEBACKER II (18–29 December 1972) comprise the bulk of the period where air strikes into North Vietnam were used in an attempt to influence the government in Hanoi, although retaliatory combat air operations into the north were conducted between the major operations.[23] As outlined, air operations into the North were initially envisioned as a graduated series of reprisal strikes, tied to specific North Vietnamese actions, and designed to lift the morale of the South Vietnamese government while simultaneously convincing the government in Hanoi to slow or eliminate support for the National

19 Ibid., x.

20 Ibid., 155. The previous September, the President had wondered aloud whether the political instability of the SVN government "made all our efforts worthless"; *Pentagon Papers*, vol. 4, sec. IV.C.3, i.

21 Ground troops were initially deployed to protect the air bases from which strikes were being flown.

22 Littauer and Uphoff, eds., *Air War in Indochina*, 9. The authors of this study had access to formerly classified documents, the *Pentagon Papers*, only for the period until the end of March 1968, later relying on publicly-released documents.

23 Under prevailing ROE, modified over time, in 1969 285 attack sorties were flown against North Vietnam, in 1970 1,113. Lewy, *America in Vietnam*, 33.

Liberation Front (NLF) in the South and, hopefully, be a catalyst for a negotiated settlement favorable to American interests.[24]

The final decision to begin "a program of measured and limited air action jointly with the GVN [Government of Vietnam (South)] against selected military targets in the DRV [Democratic Republic of Vietnam (North)]" was made by President Johnson on 13 February 1965.[25] The series of strikes was code-named Operation ROLLING THUNDER, an operation which was to continue for the next three and one-half years—far longer, and with far greater political impact, than anyone could have anticipated.

Operation ROLLING THUNDER Although the first ROLLING THUNDER strike was scheduled for 20 February 1965, a coup attempt forced its cancellation, while continued political instability in South Vietnam, and bad weather, nixed the next three planned operations. On 2 March 1965, the first successful strike, actually ROLLING THUNDER V, was conducted.[26] The operation consisted of 104 USAF aircraft attacking the Xom Bang Ammo Depot and 19 South Vietnamese aircraft attacking the Quang Khe Naval Base. The strike was deemed a success, although four aircraft were lost to enemy aircraft. As true of many military efforts, the initial focus of the operation quickly changed.

Over the next two weeks, Ambassador Taylor and General William Westmorland, Commander, U.S. Military Assistance Command, Vietnam (MACV), pressed Washington to "increase the tempo and intensity of the air strikes." The president accepted their rationale and on 15 March issued new guidelines for ROLLING THUNDER. As the authors of the *Pentagon Papers* note, the operation "had thus graduated to the status of a regular and continuing program"—a significant shift in fundamental strategy occurring in just one month's time.[27] The targeting strategy was the next key aspect of the operation to be altered.

Early target selection, for the envisioned series of discrete reprisal strikes, had been "completely dominated by political and psychological considerations."[28] However, as the overall concept of the operation shifted, there arose a need to develop a targeting concept that delivered tangible results towards the stated

24 Admiral U.S. Grant Sharp Jr., Commander-in-Chief, U.S. Pacific Command, cautioned against using the term "gradual reprisal" hoping, rather, "that we are thinking, and will act, in terms of a 'graduated pressures' philosophy which has more of a connotation of steady, relentless movement toward our objective." Department of Defense, *Pentagon Papers*, vol. 4, sec. IV.C.3, 43; Interestingly, Sharp quickly became annoyed with what he saw as the overly timid application of this strategy and, in 1978, published his book *Strategy for Defeat* (San Rafael (Ca.): Presidio Press, 1978), blaming that strategy for the American defeat. The National Liberation Front (NLF) is also referred to as the Vietcong (VC.); Pape, *Bombing to Win*, 175.

25 Ibid., iv.

26 Ibid., v.

27 Ibid., 67, 73.

28 Ibid., 74.

military objective. "The first and most obvious candidate for such a target concept was that of interdicting the flow of men and supplies into South Vietnam by striking the lines of communication (LOCs) of [North Vietnam.]"[29] While debate continued for another month, by mid-April the targeting focus had firmly shifted.[30]

From the very beginning, a series of bombing pauses were introduced during ROLLING THUNDER. McNamara hoped "[they] might trigger a sequence of actions leading Hanoi either to negotiate or to reduce its support of the insurgency, while bolstering domestic and international support for the administration's policy."[31] The first, Project MAYFLOWER, ran from 13–18 May 1965 with no apparent positive outcome, as the U.S. was "unmistakably rebuffed by Moscow, Hanoi, and Peking."[32] Such bombing pauses were to remain controversial, with many feeling that taking the pressure off only benefited the North Vietnamese war effort.[33] For example, in testimony before Congress in 1967, then-USAF Chief of Staff General John P. McConnell argued that "Portions of the labor force or defense force, if no longer required in the North due to a halt or decrease of the bombing, would be available for infiltration into the South."[34]

This short summary of the first ten weeks of Operation ROLLING THUNDER is not meant as the beginning of a detailed narrative of the extended campaign, an effort ably undertaken by many others. Rather, it is merely designed to demonstrate that ROLLING THUNDER was, from the very beginning, a complicated, dynamic, constantly changing operation. It is indeed a cautionary example demonstrating Clausewitz' dictum that "war is not an exercise of the will directed at inanimate matter, as is the case with the mechanical arts, ... [but] In war, the will is directed

---

29 Ibid.

30 Ibid., v, xix. Following high level meetings in Honolulu on 20 April 1965, the secretary of defense announced this "interdiction" policy in a special press briefing on 26 April.

31 McNamara, *Retrospect*, 185; Eight bombing pauses were ordered. Mark Clodfelter, *The Limits of Airpower: The American Bombing of North Vietnam* (New York: The Free Press, 1989), 119.

32 Department of Defense, *Pentagon Papers*, vol. 4, sec. IV.C.3, 126; Clodfelter, *Limits of Airpower*, 240n.7.

33 Even the American population was conflicted on the issue of bombing pause. In March 1968, following the Tet Offensive, one Gallup poll revealed that 51 percent of the public disagreed with a bombing pause, although those numbers dipped markedly in the next month. Clodfelter, *Limits of Airpower*, 239n.167.

34 U.S. Congress, Senate, Committee on Armed Services, Preparedness Investigating Subcommittee,

Air War Against North Vietnam, 90th Cong., 1st sess., part 3, 22–23 August 1967, 203. Later in these proceedings, Admiral Thomas H. Moorer, USN, Chief of Naval Operations concurred with this assessment, 247. It became a recurring theme among the senior military leaders called to testify; Pape argues that the senior military leadership actually argued their case well enough that the eight short pauses did not substantially hinder the interdiction efforts. Pape, *Bombing to Win*, 186–7.

at an animate object that *reacts*."[35] Simply put, despite a nation's best laid plans for achieving its political objectives, the enemy retains a vote.

As mentioned, Operation ROLLING THUNDER was to continue for more than three additional years, eventually ending on 31 October 1968. A study conducted by the Air War Study Group at Cornell University, divided the campaign into five distinct phases.[36] The first, the beginning of which was outlined above, was the ongoing campaign designed to interdict the flow of forces and materiel into South Vietnam, an effort that continued throughout the air war. The second phase of ROLLING THUNDER, which lasted one month from the end of June 1966 until the end of July, specifically targeted the destruction of North Vietnam's oil storage (POL) capacity. While the operation was initially deemed a success, as some 70 percent of POL storage capacity was destroyed, over time it was seen as a strategic failure as it did nothing to slow the flow of forces into the South.[37]

As debate continued at the highest levels of the American government, along with the ongoing realization that neither the interdiction operations nor the POL campaign had stemmed the tide of reinforcements some, including the Joint Chiefs of Staff and Admiral U.S. Grant Sharp Jr., then Commander-in-Chief, U.S. Pacific Command (CINCPAC), pushed for an escalation of air operations. This expanded program, beginning in the spring of 1967, became the third phase of operations identified by the Cornell study, and was largely conducted against targets that had previously been "off limits" or in prohibited areas.

The expanded target list included "urban power plants, a steel plant, an airfield, an urban power transformer, a petroleum storage facility, an ammunition dump, and a cement plant."[38] Expanded operations also included the mining of some rivers and estuaries and, eventually, critical targets much closer to Hanoi and Haiphong. These operations continued until the end of March 1968. The fourth phase of ROLLING THUNDER, as defined by the Air War Study Group, was the period between President Johnson's announcement of a partial bombing halt, to begin on 1 April 1968, until the end of the air war over North Vietnam on 31 October 1968.

As the study was written beginning in 1971, the authors could not benefit from a complete historical understanding of the war. As a result, what they define as the fifth phase, the continuation of interdiction operations in the North at a much reduced level, is now properly understood as the intervening period between this operation and LINEBACKER I. While the numbers are approximate, but deemed accurate, during Operation ROLLING THUNDER more than 300,000 sorties

35 Clausewitz, *On War*, 149: emphasis in original.

36 It is important to note that this study was written prior to either of the LINEBACKER operations. Thus, for these authors, ROLLING THUNDER was *the* air war over North Vietnam.

37 Littauer and Uphoff, eds., *Air War in Indochina*, 39–40; Department of Defense, *Pentagon Papers*, vol. 6, sec. IV.C.7.(a), 13–44.

38 Ibid., 40; Department of Defense, *Pentagon Papers*, vol. 6, sec. IV.C.7.(a), 13–26.

were flown into North Vietnam, delivering over 600,000 tons of air dropped munitions.[39]

Yet, in the end, this massive effort was largely seen as a failure. In the preface to a Committee on Foreign Relations staff study on the operation completed in 1972, the Chairman, Senator J.W. Fulbright, noted,

> of the five major goals set for the bombing only one has been achieved. The bombing has succeeded in making North Vietnam pay a high price for her support of the war in the South. But the air war has not stopped the flow of supplies to the South, broken Hanoi's will, or forced the North Vietnamese to negotiate an end to the war.[40]

With regards to PGMs, however, the ROLLING THUNDER period from 1965–1968 is quite interesting.

In April 1965, at very nearly the same time the operation was commencing in Southeast Asia, Colonel Joseph Davis, USAF, head of the Directorate of Technical Assistance and Support was in Orlando, FL being briefed on laser seekers.[41] Recognizing the possible benefits of the application of this technology to air-delivered munitions, Davis successfully pressed forward with the concept and by June permission to award two feasibility prototype contracts was obtained. Flight tests conducted by the two companies selected, Texas Instruments and the Autonetics Division of North American Aviation, were both successful, with the engineering prototype contract being awarded to Texas Instruments on 20 May 1967.[42]

While the engineering prototype test series began in November 1967 at Eglin AFB in Florida, they were quickly moved to Southeast Asia where the Operational Test and Evaluation (OT&E) was performed between May and August of 1968.[43] However, although LGBs were introduced into the Southeast Asian theater prior to the bombing halt, they were only employed against the southernmost portion of North Vietnam until later in the war.[44] Meanwhile, as the Air Force was developing

---

39 Ibid., 168–70. Fig. 13-2 provides the aerial munitions tonnage numbers, while figs. 13-3 and 13-4 provide aircraft sortie information. These sortie numbers include both tactical aircraft and B-52s. Other sources quote numbers as high as 864,000. See Department of Defense, *Pentagon Papers*, vol. 6, sec. IV.C.7.(b), 114. For consistency across sub-theaters, the Cornell study numbers will be used.

40 U.S. Congress, Senate, Committee on Foreign Relations, *Bombing as a Policy Tool in Vietnam: Effectiveness*, 92nd Cong., 2nd. sess., Study no. 5, 12 October 1972, v.

41 Peter DeLeon, *The Laser-Guided Bomb: Case History of a Development*, Rand no. 1312-1 (Santa Monica, CA, 1974), 8–9.

42 Ibid., 10–17. DeLeon notes the accelerated procurement process slowed down once the concept was validated, in keeping with standard Air Force policies.

43 Ibid., 24–6.

44 William W. Momyer, *Air Power in Three Wars*, 1979, Reprint (Maxwell AFB, AL: Air University Press, 2003), 238–9; Blackwelder, *Long Road*, 14–16; Werrell, *Chasing the*

its laser-guided capability, the U.S. Navy was leading the way in another type of PGM.

The Navy's efforts in electro-optical guidance lead to the development and introduction of the Walleye I, a free-falling hybrid of earlier glide bomb and vertical bomb technologies. This early version was a "fire and forget" weapon. In simple terms, the pilot, using the bomb's own television seeker, "locked" the weapon onto the target. Following release the weapon used a contrast seeker system to fly to the target. With its linear-shaped-charged warhead, Walleye I was found to be very effective against bunkers, barracks, and smaller bridges.

Against larger bridge abutments, however, the 850-pound warhead was less effective. For example, on 12 March 1967 three Walleyes were used against the Than Hoa Bridge ("The Dragon's Jaw") located 75 miles south of Hanoi.[45] All the weapons hit near the aim point, but with no discernible effect. In 1972, Walleye II, with a larger 2000-pound warhead, was introduced with better operational results.[46] As the war progressed, the two services traded technologies, with each employing both LGBs and EO weapons, although far fewer of the latter were used during the war.[47] With few PGMs being used in North Vietnam during this period, ROLLING THUNDER provides a useful baseline, as the final "non-precision era," against which future operations may be compared.

Operation LINEBACKER I As long as the character of the war in Vietnam primarily remained that of an insurgency, air power was to have limited success in bringing the conflict to an end on terms favorable to the U.S. and South Vietnamese governments. But by 1972, the political objectives of the U.S. and the military strategy used by the government in Hanoi had both changed, thereby opening up an opportunity for the U.S. to wield air power in a much more effective fashion. As Robert Pape writes, "Although the initial [ROLLING THUNDER] campaign failed, the Linebacker I and II campaigns succeeded in compelling Hanoi to sign the 27 January 1973 Paris agreements, ending America's role in the war."[48] But how did these two changes alter the fundamental dynamics present, and what role, if any, did PGMs play in the final outcome?

First, it is important to understand a key American policy objective. From at least as early as November 1963, the senior leadership in Washington knew that while the U.S. would help South Vietnam in its struggle, it would eventually be

---

Silver Bullet, 149. While Momyer states than no LGBs were used in North Vietnam before the bombing halt, Blackwelder and Werrell document that the OT&E actually occurred in Route Package 1, the southernmost part of North Vietnam.

45 Werrell, *Chasing the Silver Bullet*, 151.

46 Blackwelder, *Long Road*, 17–18; Gillespie, *Precision Guided Munitions*, 153.

47 Ibid., 19. The first use of Walleye by the Air Force was on 24 August 1967. Cost ever the factor, EO weapons were more expensive than LGBs.

48 Robert A[nthony] Pape, *Bombing to Win: Air Power and Coercion in War* (Ithaca, NY: Cornell University Press, 1996), 195.

up to the South Vietnamese to defend themselves against the North. As President Kennedy remarked, "In the final analysis it is their war. They are the ones who have to win or lose it."[49] Despite this early acknowledgment, for more than two years the U.S. carried the majority of the offensive load as the South Vietnamese were seen as unable to carry out the task.

This began to change in mid-1967 with a process that became known as "Vietnamization," an effort which accelerated under President Nixon following the 1968 general election. While there were certainly many difficulties in training and equipping the armed forces of South Vietnam, by late 1971 "Vietnamization seemed to be working. ARVN forces, wrote a highly placed MACV officer ... , 'have demonstrated their ability to work without U.S. advisory assistance and have done remarkably well. Military competence was high and the future could be faced 'with confidence.'"[50]

As a result of this perceived success, the American government moved to a more narrowly focused, but clearly defined political objective. No longer required to provide the preponderance of forces to help win the war for the South Vietnamese, the Nixon Administration merely wished to conclude a final negotiated settlement with North Vietnam, through the often difficult Paris Peace Accord process.

The second significant change is to be found in the nature of North Vietnamese operations at this juncture of the war. Having determined that their prospects were not good in the South, where Vietnamization and Pacification seemed to be working, "the decision was made to abandon the strategy of protracted war and to force a decision by conventional invasion."[51] Early in the morning of 30 March 1972, the North Vietnamese launched a large-scale conventional force attack into South Vietnam, an action soon known as the Easter Offensive. Some South Vietnamese forces acquitted themselves well, while others performed less admirably.[52] But one major factor contributing to the failure of this North Vietnamese campaign over the next seven months was the difference that American air power made.

In a series of rapid escalations, President Nixon authorized resumed bombing of North Vietnam beginning on 6 April.[53] By 10 May, the day Operation LINEBACKER began, air operations were authorized

> throughout North Vietnam except in the 25–30-mile-wide buffer zone alongside the Chinese border and the restricted areas 10 miles from the centers of Hanoi and Haiphong where specific strike authorization was required. At the same time

49 Quoted in Lewy, *America in Vietnam*, 162.

50 Ibid., 167.

51 Ibid., 196.

52 For short accounts of the Easter Offensive see Lewy, *America in Vietnam*, 196–201; Pape, *Bombing to Win*, 195–201.

53 Operations initially limited to within 25 miles north of the DMZ, were quickly expanded to the 19th then 20th parallels.

> the president ordered the mining of North Vietnamese harbors and the blockade of the coast, and action for which the JCS had pleaded since 1965.[54]

What made this operation so effective, especially when compared to the extended ROLLING THUNDER campaign, was that the North's tactical *modus operandi* now played directly to American air power's strengths, particularly as the force was now armed with new, far more accurate munitions.

Before, the dispersed nature of insurgent forces meant they were difficult to target effectively. Now, massed conventional forces, including trucks and tanks, were prime targets for air power. Additionally, the wider availability and use of precision-guided munitions meant that once difficult targets were readily destroyed. The initial strike of the campaign involved some 90 aircraft. Due to the accuracy of the PGMs, only about one-third of these were "strike" aircraft, with the remainder flying in support roles. During the attack, PGMs were used to heavily damage Hanoi's Paul Doumer Bridge, while conventional ordnance was used against a rail yard.[55]

In another well-known example, the Than Hoa Bridge south of Hanoi which, since 3 April 1965, had survived 871 sorties flown against it during ROLLING THUNDER including the unsuccessful Walleye I attacks in March 1967, was felled with laser-guided bombs dropped from four aircraft on 13 May.[56] On 26 May three aircraft destroyed three buildings in the Son Tay warehouse and storage area, each almost the size of a football field. Until August, this was how most strikes were conducted in the North, with small numbers of strike aircraft, using PGMs, being accompanied by larger support packages. The success of missions such as these led Major General Eugene L. Hudson, 7th Air Force Director of Intelligence, to observe that, "Laser-guided bombs ... revolutionized tactical bombing."[57] Perhaps more importantly, although not to be realized until much later, the use of such weapons was to begin to influence the thinking of the nation's senior military and political leaders.

In August, air operations intensified as President Nixon sought to place increased pressure on the North Vietnamese in advance of the next round of negotiations in Paris. Weather often was a factor as LGBs required a clear line of sight from the attacking aircraft to the target to guide properly. September was a

54 Lewy, *America in Vietnam*, 410.

55 Clodfelter, *Limits of Airpower*, 158–9. The operation was initially called "ROLLING THUNDER ALPHA," as the code name LINEBACKER "had not yet reached field units."

56 Edgar Elsamer, "The Steady Evolution in Armaments," *Air Force Magazine* 68.12 (December 1985): 79. Additionally, eleven American aircraft had been lost on the earlier missions, none were lost this day; Lewy, *America in Vietnam*, 410–11; Anderegg, C.R., *Sierra Hotel: Flying Air Force Fighters in the Decade After Vietnam* (Washington, DC: U.S. Government Printing Office, 2001), 122.

57 Quoted in Clodfelter, *Limits of Airpower*, 159.

particularly busy month, with four times as many strike sorties being flown when compared with previous months. Planners continued to use precision-guided munitions as much as possible, although there were never enough guidance pods for all missions. Despite this, the 7th Air Force flew over a hundred LGB-armed sorties in September, most of which destroyed their targets.[58]

As a result of progress at the negotiating table, Nixon ended Operation LINEBACKER on 23 October 1972, by suspending bombing above the 20th parallel. During this seven-month period, over 150,000 tons of bombs were dropped, roughly a quarter of the amount delivered during the three-and-a-half year campaign of ROLLING THUNDER, with PGMs inflicting a majority of the damage.[59] The air war in the North, however, was not over yet.

Operation LINEBACKER II While the initial LINEBACKER campaign was able to wring concessions from the North Vietnamese, it was not enough to bring a successful conclusion to the negotiations. South Vietnam's President Nguyen Van Thieu had earlier introduced modifications to the 8 May peace proposal and the two sides had difficulty coming to agreement. Certain that Congress was going to end funding for the war upon its return in January 1973, Nixon was willing to risk the negative publicity increased military action would likely bring. In early November he once again authorized B-52 flights north of the de-militarized zone (DMZ). The gambit worked.[60]

Yet when the parties returned to the negotiating table in Paris on 20 November, "[National Security Advisor] Kissinger noted that his adversary was not the same [North Vietnamese negotiator] Le Duc Tho of late summer who relentlessly pushed towards a settlement."[61] Difficult, on-again, off-again, negotiations continued into mid-December. Meanwhile, planners worked on a potential military response if the negotiations failed. Finally the president authorized a three-day bombing campaign against Hanoi on 14 December.[62] Operation LINEBACKER II was soon underway.

On the evening of 18 December, a total of 129 B-52s in three waves, struck targets in and around Hanoi in the first of what became an eleven-day campaign.[63] By design, the air campaign was meant to cause distress to the civilian population, without actually killing it. In doing so, planners necessarily walked a fine line. "While seeking to avoid civilian casualties, air chiefs complied with Nixon's

58 Ibid., 161.

59 Ibid., 166–7.

60 Ibid., 179; Pape, *Bombing to Win*, 176.

61 Ibid.

62 Lieutenant General Gerald W. Johnson, USAF, Eighth Air Force Commander on Guam, found out on the 15th. James R. McCarthy and George B. Allison, *Linebacker II: A View from the Rock* (Maxwell AFB, AL: Air War College, 1979), 39.

63 Ibid., 1; Clodfelter, *Limits of Airpower*, 185–6, 189.

desires and designed Linebacker II to inflict the utmost civilian distress."[64] In keeping with the delivery platform/weapon allocation strategy developed during the first LINEBACKER campaign, B-52s attacked larger area targets, generally on Hanoi's periphery, while U.S. Navy and Air Force tactical fighters, using PGMs, struck more difficult targets in the more heavily populated areas of the city.[65]

During the eleven-day campaign, 729 B-52 sorties were flown dropping more than 15,000 tons of munitions. Tactical aircraft flew an additional 1216 sorties dropping about 5,000 tons of munitions.[66] Despite this massive effort so close to population centers, the operation resulted in remarkably few civilian casualties. "Hanoi's mayor claimed 1,318 civilians killed and 1,216 wounded, while Haiphong reported 305 dead."[67] This final bombing campaign broke the stalemate and the cease-fire agreement was signed in Paris on 27 January 1973.

Aerial Operations in the South

As earlier noted, air operations in support of the government in Saigon against the NLF began as early as 1962. Over the next three years, more than 14,000 sorties were flown by American personnel.[68] However, the scope of these operations pale in comparison to the level of air operations conducted once American involvement in the war expanded after the Tonkin Gulf Resolution was passed. Between 1965 and 1971, more than 950,000 fixed-wing sorties were flown in South Vietnam, delivering roughly 3,900,000 tons of aerial munitions.[69] While the air war in North Vietnam may have garnered much of the attention on the world stage, particularly between 1965 and 1968 and again in 1972, the preponderance of the air effort was clearly in the South throughout the war.

Defining the character of air operations in the South is, simultaneously, easier and more difficult than of those in the North. The Cornell study's analysis indicates that less than 10 percent of the air effort by fixed wing aircraft involved direct support to friendly forces engaged in battle.[70] Thus, roughly 90 percent of flights were broadly categorized as "interdiction" missions.[71] Yet this simple characterization masks a much more complicated picture. Such "wide and diffuse" operations can, and did, include interdiction of LOCs, harassment operations,

---

64 Clodfelter, *Limits of Airpower*, 184.

65 Ibid., 184; Momyer, *Three Wars*, 266.

66 Ibid., 194.

67 Clodfelter, *Limits of Airpower*, 195.

68 Lewy, *America in Vietnam*, 24, Table 1-1.

69 Littauer and Uphoff, eds., *Air War in Indochina*, 274–7, Tables SS-5a.(i) and SS-6, 11, fig. 1-1.

70 Ibid., 10, 53.

71 Ibid., 55. "Interdiction [being] defined as preventing or hindering enemy use of an area or route, or as a selective campaign against the enemy's total logistic organization."

reprisal strikes, and even tactical air strikes into cities.[72] Significantly, even the geo-strategic situation was more complicated in the South than in the North.

While operations into North Vietnam can be characterized as being launched "into enemy territory," operations into the South, much larger in scale, were being conducted in the very country being assisted by the American forces. As the study notes:

> Because of this wide range of targets and objectives, no unique characterization of the bombing in South Vietnam is possible. However, it can clearly not be considered to have been restrained or highly discriminating, if only because of the sheer volume of activity; and, though the strikes may have been motivated by military objectives, the intermingling of combatants and civilians in South Vietnam results in severe social and political consequences from the massive deployment of air power.[73]

From within this short excerpt, a painful realization emerges.

It is axiomatic that the deaths of, and/or injury to, innocent non-combatants are regrettable in every instance, even if they might be seen as morally acceptable under the just war principle of double effect. But the deaths of the citizens of the nation you are trying to protect, *within their own borders*, only serves to obfuscate and complicate the role of the supposed foreign ally. As Lewy observes, "It was difficult to convince villagers that the Americans had come as their liberators if in the process of liberating them from the Communists allied troops caused extensive harm to Vietnamese civilian life and property."[74]

One of the key challenges of air operations in the South, which was to cause so much suffering in that country, was the great difficulty that U.S. and South Vietnamese forces had in precisely locating the enemy. Often commingled with other Vietnamese, and able to move quickly and agilely through the densely vegetated countryside, NLF guerilla forces were, in many if not most instances, invisible from the air.

Unable to locate and precisely target such forces, the U.S. resorted to using either area coverage weapons, such as napalm or cluster-bomb units (CBUs), or area bombing techniques.[75] The Cornell study, for example, calculates that "more than half of the tonnage of aerial munitions dropped on South Vietnam was delivered by B-52s."[76] Being subjected to carpet bombing by B-52 aircraft certainly had a negative psychological effect on the enemy forces, an effect that was purely intentional. As one senior Vietcong official recalled, "The first few

72 Ibid., 10–11, 55.

73 Ibid., 11.

74 Lewy, *America in Vietnam*, 95.

75 Cluster bombs are canister-type weapons that contain anti-personnel and anti-equipment sub-munitions, which disperse over a wide area when delivered.

76 Littauer and Uphoff, eds., *Air War in Indochina*, 57.

times I experienced a B-52 attack it seemed, as I strained to press myself into the bunker floor, that I had been caught in the Apocalypse. The terror was complete. One lost control of bodily functions as the mind screamed incomprehensible orders to get out."[77]

However, it is difficult to view such bombing as particularly proportional or discriminate in just war terms. As one author of the Cornell study records:

> The B-52 Stratofortress was originally meant to deliver the atom bomb, but its enormous cargo capacity suggested a new concept in the conduct of the Indochina air war. Outfitted with racks to carry conventional bombs, a B-52 can deliver up to thirty tons of explosives in a dense pattern. A typical mission, consisting of six of these bombers flying in formation, saturates an area of more than a square mile with bombs. Everything within this area is demolished. To witness such a raid is to witness a disaster of major proportions.[78]

While concerted efforts were made to clear areas of civilians before many such operations commenced, suffering by non-combatants in the South was significant. "For the period from 1965 to April 1971, the estimate of civilian casualties in South Vietnam is 1,050,000 including 325,000 deaths, while over 6,000,000 South Vietnamese (one-third of the population) are thought to have become refugees."[79] As was the case elsewhere, as PGMs became more readily available their use expanded in operations in the South. One two-week period in November 1972, for example, saw almost 500 LGBs released by U.S. Air Force pilots on armed reconnaissance missions.[80] Electro-optically-guided weapons also were in use, with 134 being released in one two-month period between 15 September and 17 November 1972.[81]

Aerial Operations in Laos and Cambodia

Combat missions were also flown in the neighboring countries of Cambodia and Laos. One of the primary tasks for many of these missions was to interdict the so-called "Ho Chi Minh Trail." This so-called "trail," actually a diffuse network of roads and paths, served as a primary logistics route from North Vietnam into the South. Supplies and personnel flowed from North Vietnam, through mountain passes into eastern Laos, then south and east, in generally heavy jungle and along

77 Vietcong Minister of Justice Truong Nhu Tang, quoted in Clodfelter, *Limits of Airpower*, 183.

78 Littauer and Uphoff, eds., *Air War in Indochina*, 56.

79 Ibid., 63.

80 Roy E. Metcalf, "Guided TACAIR Weapons in Southeast Asia," Scientific Advisory Group Working Paper No. 4-72 (Department of Defense, U.S. Pacific Command, 28 November, 1972), Slide 4, 9. Weapons data is for USAF-released weapons only.

81 Ibid., Slide 6, 13.

rough terrain, into South Vietnam. Other branches continued further south into Cambodia, then east into the South. Combat missions were also flown in support of the allied governments in both countries. For many years, for example, missions were flown against the Pathet Lao in northern Laos and, later, against Khmer insurgent forces seeking to overthrow the government of Lon Nol in Cambodia.

Operations in Southern Laos American air interdiction of communist supplies along the Ho Chi Minh Trail in Southern Laos began in 1965. Under the codename STEEL TIGER, which was later subsumed under Operation COMMANDO HUNT in late-1968, operations were to continue for seven years. At first the operation was relatively modest, with an average of less than 50 tactical aircraft sorties per day. However, as the Cornell study notes, "There was a close relationship between Trail interdiction and the air war against North Vietnam ... The two operations competed for aircraft, and planes which were unable to strike their scheduled targets in North Vietnam were often diverted onto the Trail in Laos."[82] After the bombing halt in the North, efforts along the Trail more than doubled for the remainder of the war, hovering between roughly 180 to 225 sorties per day, with more than 380,000 tactical sorties eventually being flown.[83]

From the beginning, interdiction operations proved challenging. The difficult terrain, thick jungle, and dispersed nature of both the Trail and those using it made effective operations elusive. One solution was to use the wide area bombing capability of the B-52 to render the trail impassable. In 1971, for example, about 1,000 B-52 sorties per month were being flown against the Trail.[84] In addition to this raw firepower-based solution, technological innovations such as low light-level television systems, air-dropped sensors, and infra-red systems were introduced, becoming other early entrants into what was to become the complex electronic battlefield of today.[85]

The amount of ordnance delivered in these interdiction operations steadily increased between 1965 and 1971. In the end, more than 1,100,000 tons of ordnance was delivered along the trail. Because of a relatively low population density, civilian suffering in this area was relatively mild, although there was, by one estimate, internal displacement of roughly 5 percent of the local population.[86]

Along with the increase in total tonnage of ordnance dropped, newer PGMs also played a role in this particular effort. In fact, the first LGB dropped in combat, on

82 Littauer and Uphoff, eds., *Air War in Indochina*, 69.

83 Ibid., 69–70. Daily averages in Figure 5-2, 70. Detailed tactical sortie data in Table SS-5a.(iii), 275.

84 Ibid., 70; Momyer notes that LGBs were used in Laos starting in 1968. Momyer, *Three Wars*, 239.

85 Ibid; These are some of the additional types of technologies to which Lacquement refers. See Lacquement, *Shaping*, 23.

86 Ibid., 11, 75.

23 May 1968, was dropped on just such an interdiction mission.[87] One U.S. Air Force working paper records that during the period August 1969 to February 1970 in Southern Laos, 1263 LGBs were released, with largely positive results.[88] During a later period, April/May 1971, both laser and EO-guided MK-84 (2,000 lb.) bombs were employed, directly striking their targets greater than 50 percent of the time.[89] In a significant tactical development, during the spring of 1971 LGBs were used for the first time against tanks in combat, with positive results.[90]

Operations in Northern Laos Whereas American air operations in the southern portion of Laos were principally focused on interdicting communist logistics along the Ho Chi Minh Trail, operations in the northern part of the country were of an entirely different nature. There, American forces flew missions in support of the Royal Laotian government against Pathet Lao insurgents, and their North Vietnamese supporters. While direct American involvement began in the spring of 1964, by the spring of 1965 the U.S. operation in Laos was divided into two programs, with Operation BARREL ROLL, in the north, joining STEEL TIGER to the south.[91]

For the first four years of the operation, the daily air activity by tactical aircraft was quite low, averaging well below 50 sorties per day. As was the case elsewhere though, the cessation of bombing in North Vietnam freed aircraft for tasking. Reacting to a sustained, and much larger, Pathet Lao/North Vietnamese offensive in 1969, tactical aircraft sorties rose dramatically, eventually topping out at around 300 per day. Sortie rates began to drop in 1970 and by 1971 the steady state effort hovered at just below 100 sorties per day.[92] This apparent drop off in level of effort, however, masked a more complicated situation, as in February of 1970 President Nixon authorized the use of B-52 missions into Northern Laos, although the publicly announced scope of the mission was nuanced at best.[93] The steady use of B-52s into Laos continued unabated from that point forward, and by the end of 1971 more than 500,000 tons of munitions, from all platforms, had been dropped.[94]

Although acknowledging that "In many areas threatened by a PL/NVA advance the population has been evacuated, leaving the area a free-fire zone," the Cornell

87 Anderegg, *Sierra Hotel*, 124. That honor went to the "Wolfpack" of the 8th Tactical Fighter Wing flying out of Ubon Royal Thai Air Base.

88 Metcalf, *Guided TACAIR Weapons*, Slides 1, 3.

89 Ibid., Slide 2, 5.

90 Momyer, *Three Wars*, 362.

91 Littauer and Uphoff, eds., *Air War in Indochina*, 76–7.

92 Ibid., Table SS-5a.(iii), 275. Between 1968 and 1971 143,000 tactical aircraft sorties were flown.

93 Ibid., 78–9. Figure 6-1 documents the estimated averages from 1965–1971.

94 Ibid., Figure 1-1, 11. This figure includes ordnance dropped by both tactical aircraft and B-52s.

study documents that the costs imposed on the people of Laos as a result of this portion of the war in Southeast Asia were tremendous.[95] It observes:

> Western journalists who have visited the Pathet Lao headquarters in the province of Sam Neua report a landscape covered with bomb craters, 'a chaos of red earth, broken rocks, devastated trees.' Most towns in the area have been destroyed. The villagers, and Pathet Lao live in trenches and caves; farming, when possible at all is done at night. Interviews with refugees from the Plain of Jars indicate similar living conditions there. ... Similar observations of the destruction in Laos have been reported by virtually all those who have had an opportunity to visit the regions involved.[96]

Such a level of devastation is difficult to imagine for those who have not seen it.

Operations in Cambodia Until April 1970, American combat air operations in Cambodia were extremely limited, with occasional sorties being flown against North Vietnamese forces along the frontier with South Vietnam. This changed dramatically after the invasion of Cambodia by American and South Vietnamese forces following the fall of the Sihanouk government that month.[97] During 1970 and 1971, American aircraft, tactical and B-52s, flew a full range of combat missions dropping roughly 200,000 tons of munitions.

Even later, in March and April of 1973, American combat sorties in Cambodia were still being flown on two distinct missions. In March, for example, roughly 20 percent of sorties were flown against Khmer insurgent forces in support of the Cambodian government headed by Lon Nol. The remaining 80 percent were interdiction missions being flown against North Vietnamese supply lines into the South. Interestingly, by mid-April, the percentages had reversed.[98]

Insurgent forces proved as difficult to locate in Cambodia as elsewhere, thus interdiction operations were as relatively ineffective here as in other sub-theaters of the war. The impact on the civilian population was also similar. A December 1971 *Washington Post* article reported that those towns and villages taken over by the insurgents were being subjected to extensive bombing.[99] A 1972 GAO report on refugees in Cambodia reported:

---

95 Ibid., 15.

96 Ibid., 81–2.

97 Ibid., 15, 87. Table SS-5a.(v), 275.

98 U.S. Congress, Senate, Committee on Foreign Relations, Subcommittee on U.S. Security Agreements and Commitments Abroad, *U.S. Air Operations in Cambodia: April 1973*, 93rd Cong., 1st sess., 27 April 1973, 1, 8.

99 Noted in Littauer and Uphoff, eds., *Air War in Indochina*, 89. Original article in December 17, 1971 edition of the *Washington Post*.

> More than two million Cambodians have been driven from their homes at different times since the war spread to their country in the spring of 1970. ... Bombing is a very significant cause of refugees and civilian casualties. ... Refugee conditions are deplorable and hospitals are suffering from severe pharmaceutical shortages. ... The population of Phnom Penh is estimated to have more than doubled because of the influx of refugees.[100]

Observations such as this, coupled with the earlier reports from Laos, demonstrate the severe level of destruction that was often present prior to the large scale use of PGMs.

An Analysis of the Performance of Precision-Guided Munitions

There are a number of distinct challenges when analyzing the performance and effect of precision-guided munitions in the Vietnam War. The first is that, as opposed to the Second World War, and the later Gulf War of 1990–1991, there was no comprehensive post-war study of the air war in Southeast Asia.[101] There simply is no convenient single source to turn to for complete and definitive statistics and analysis; as a result, the data is fragmented. However, there is more than enough raw data embedded in the various primary and secondary sources available to allow the study to begin to draw some broad conclusions.

The second is that, chronologically, the Vietnam War spans the dividing line between the era before precision-guided munitions existed as a truly viable technology and the era when they are increasingly viewed as a potent and war-winning alternative. Thus, the hybrid nature of the war makes it difficult, along some lines of inquiry, to draw clear distinctions between the performance and effects of precision weapons versus their earlier non-precision antecedents. As the Cornell study concedes, "In assessing the record of past actions, it is often impossible to separate the effects of air warfare [or, in this case, precision-guided air warfare] from the rest of the fighting [or non-precision air warfare.] In this area one must be content with finding mixed causes and mixed consequences." [102]

Next, and this is similarly true for the project as a whole, each case study may not prove to be a rich source of evidence along all seven of the "lines of inquiry" embedded in the three-part evaluative framework. While some cases may provide ample evidence across the breadth of the framework, in others it is possible that data relevant to only a few lines of inquiry might be present. This realization, however, only serves to highlight the value of selecting a number of diverse case studies to examine.

---

100 Ibid., 89–90.

101 U.S. Strategic Bombing Surveys for the Pacific and European theaters were published in 1947. The Gulf War Air Power Survey was commissioned 1993.

102 Littauer and Uphoff, eds., *Air War in Indochina*, xxi.

It must be reemphasized that this study is not analyzing the perceived, or actual, overall political or military effectiveness of the various bombing campaigns, or the moral correctness of those actions, or indeed of the wider war itself. The goal here is to begin to understand the impact that this particular technology has had on human suffering in conflict, and on how this capability has altered the decision-making process of those who choose to employ military force to further national objectives.

Despite these challenges, sufficient evidence does exist to enable the study to usefully begin its longitudinal analysis. In fact, the most meaningful outcome of the analysis of this case study may lie in the development of a baseline understanding of the impact conventional air dropped ordnance had on human suffering in this conflict. Emerging from that knowledge, coupled with a basic understanding of how precision-guided weapons have fundamentally altered this baseline, identifiable and useful trends may begin to be identified.

Part I: The Weapons and Their Employment

As outlined previously, this first level of analysis examines PGMs as a technology, a new and novel class of weapons in and of themselves, trying to determine if they are indeed as accurate and reliable as is touted, and to examine if they are, in practice, being employed in a manner designed to reduce human suffering. With this case, beginning as it does before more modern PGMs were available in the inventory, the study has an excellent opportunity to directly compare the employment of PGMs against the employment of earlier "dumb" weapons in the same operational and geo-strategic environment.

As noted, combat CEPs in earlier conflicts were measured not in the tens of feet, but in the hundreds and thousands of feet. This was also broadly true in the early days of the Vietnam War. Barry Miller, writing for *Aviation Week & Space Technology*, records that:

> When American bombing began in Vietnam, the Air Force cited in Congressional testimony its ability to achieve CEPs of 750 feet. Only two or three years ago pilots of Republic F-105 fighter-bombers were achieving CEPs in Vietnam of 365 feet against undefended targets; 540 feet against moderately defended targets, and 575 feet against highly defended targets. ... Today, these figures in combat are roughly 250 feet.[103]

To compensate for this operational limitation, pilots and planners were forced to employ mass to overcome the inherent inaccuracy of the prevailing aircraft delivery systems. Two of the most common ways were to drop multiple weapons from one aircraft, in an attempt to raise the probability that one or more weapons would impact the target, or to employ multiple aircraft against a single target.

103 Ibid., 24n.27.

However, even these tried and true techniques did not guarantee success. For example, and as a point for later comparison, the two FLAMING DART missions in February 1965 used a total of 267 sorties (including support aircraft) to attack 491 buildings, of which only 47 were destroyed and an additional 22 damaged.[104]

During this period it was also often suggested that the B-52, with its advanced radar and weapons system was capable of "surgical bombing," particularly when striking large, area targets. However, as Chester Cooper records, as Operation ROLLING THUNDER was coming to an end:

> The President agreed to postpone his decision to limit bombing until the [Joint] Chiefs [of Staff] had been given their opportunity to 'take out' Hanoi's thermal power plant. It would be a simple matter that could be executed with 'surgical precision'; one strike would do the job. ... On May 19th [1967], the Joint Chiefs were given their chance to get the last piece of candy out of the box. The power plant was bombed. But the target was not destroyed with 'one strike.' Nor was the raid characterized by the precision that had been advertised. Many more attacks over many more months were necessary before it was put out of operation; in the process the surrounding area took a beating. The power plant was located in downtown Hanoi and was bounded on two sides by residential areas. 'Subsidiary damage' was significant, and it was another case of 'sorry about that.'"[105]

So, against this broad understanding, how did precision-guided munitions perform during the Vietnam War?

Operational Effectiveness One of the most highly touted attributes of precision-guided weapons is their exceptional accuracy. If this claim is true, then it would enable aircrew to prosecute a target with both a high degree of discrimination and using a proportionately lower amount of ordnance, which should ideally result in a reduced amount of collateral damage. The first key line of inquiry is whether this is actually true or, perhaps more correctly, was it true for this particular case study? The evidence would suggest that, even at this early stage of their development and employment, the answer is clearly yes.

During the initial LGB test drops by North American-Autonetics and Texas Instruments, once some very early design flaws were corrected, CEPs rapidly fell below 30 feet despite less than 10 of this new breed of weapons having been dropped by either company.[106] This trend continued as the weapons underwent further testing and eventual use in combat. One formerly classified study,

104 Department of Defense, *Pentagon Papers*, vol. 4, sec. IV.C.3, 64. These results did not impress McNamara who questioned CJCS on this outcome.

105 Chester L. Cooper, *The Lost Crusade: America in Vietnam* (New York: Dodd, Mead & Co., 1970), 375–6.

106 DeLeon, *Laser-Guided Bomb*, Table 2, 14.

undertaken by the U.S. Pacific Command's Scientific Advisory Group, analyzed a portion of the LASER and electro-optically (EO) guided weapons dropped by the U.S. Air Force between the fall of 1969 and November 1972.[107] It is important to note that the EO weapons referenced in this particular study were not the Navy-developed Walleyes discussed earlier, but later USAF-modified MK-84 bombs with an electro-optical guidance kit installed. This gave the weapon, nicknamed Hobo (*Ho*ming *Bo*mb), a Walleye-like launch and leave capability.[108]

One portion of the study, examining a seven-month period of interdiction operations in southern Laos (COMMANDO HUNT III/STEEL TIGER), reveals that of 1263 LGBs released, 673 (53.3 percent) were observed to be direct hits. The observed CEP during the period hovered around 10 feet, far below the 250–400 foot CEPs being observed for conventional ordnance.[109] During another period in southern Laos, April/May 1971 (COMMANDO HUNT III/STEEL TIGER), against a variety of targets, both LASER and EO-guided weapons were observed as direct hits between 49 and 64 percent of the time.[110]

While these two parts of the study did not explain why some weapons failed to strike on/near their designated target, an analysis of a short period in November of 1972 offers additional insight. In 17 days of operations 496 LGBs were released, striking their targets 48 percent of the time, which is consistent with the previous data. But this section's expanded analysis also records that 20 (4 percent) of the weapons failed to explode (i.e. were "duds"), 44 (8.9 percent) failed to guide properly, and the impacts of 35 (7 percent) were not directly observed and therefore, under the study's strict methodology, could not be counted as successful.[111] A strike into North Vietnam on 22 May 1972 also demonstrated the advantages gained by using PGMs. "Eight F-4s carrying 16 laser bombs destroyed five bridges and damaged a sixth. During the 1965–1968 [ROLLING THUNDER] campaign, such destruction would have required a much larger number of sorties."[112]

EO-guided weapons also consistently achieved impressive results during combat operations. Evaluating Walleye for a ten-week period in 1967, the Air Force evaluation concluded that "when all went favorably the Walleye/F-4 combination was a devastatingly effective weapons system" achieving "pinpoint accuracy well within the stated 15-foot CEP."[113] A detailed breakdown of Hobo EO-guided weapons in the Pacific Command's study, for a two-month period in the fall of 1972, reveals a similar success rate, 49 percent were direct hits, although higher

107 Metcalf, *Guided TACAIR Weapons*, 1.
108 Gillespie, *Precision Guided Munitions*, 166.
109 Metcalf, *Guided TACAIR Weapons*, Slide 1, 3.
110 Ibid., Slide 2, 5.
111 Ibid., Slide 4, 9.
112 Momyer, *Three Wars*, 266.
113 Gillespie, *Precision Guided Munitions*, 155.

rates of duds and no-guides (8.2 and 17.9 percent respectively) were observed.[114] The study also expands the analysis in other useful directions.

Against a number of standard target types routinely engaged during the war, the study analyzed the observed relative advantage gained by using PGMs rather than general purpose gravity bombs. These "relative increases in sortie effectiveness" ranged from a factor of just under three to a factor of ten.[115] Simply put, the increase in sortie effectiveness meant fewer aircraft had to fly over the target in enemy threat envelopes, and less errant ordnance was landing away from the designated target. General William Momyer, then head of the Tactical Air Command (TAC), records that in 1972 the combat CEP of LGBs was "approximately 30 feet, as compared to the dive bomb CEP of 420 feet in the 1960–68 period."[116] The results detailed in the examples above closely track with that of other sources that documented and analyzed the use of thousands of USAF laser-guided bombs during the conflict.[117]

The evidence is therefore convincing that PGMs, even in early combat use in Vietnam, successfully guided and were far more accurate than their conventional counterparts. In the Indochina Theater of war, across time, geographic location, and combat mission, LGBs and EO weapons were functioning as designed approximately 75 percent of the time and achieving CEPs inside of 30 feet, with more than half observed as direct hits.[118] With greater accuracy came the ability to prosecute targets with greater discrimination, thereby allowing fewer sorties, each dropping fewer bombs, to be flown. An added benefit was that fewer attacking aircraft were being destroyed. As Paul Gillespie noted, the fact that "[n]o [USAF] Paveway aircraft were lost during the early years of implementation was a boon not only for the aircrews involved, but also for military and government leaders faced with an increasingly unpopular war."[119] It should be noted that not all support aircraft were so fortunate.

Logically, the benefits resulting from this enhanced capability should also have resulted in lower levels of collateral damage being experienced by those on the ground, a point which will be expanded upon in the next section. But against the framework's two metrics for this line of inquiry, CEP and Weapons Functional Effectiveness, the trend lines for the PGMs are strongly positive, even at this early stage of operational employment.

---

114 Metcalf, *Guided TACAIR Weapons*, Slide 6, 13.

115 Ibid., Slide 3, 10. These numbers remained consistent across two other data sets examined. See Slide 8, 17 and Slide 9, 19.

116 Momyer, *Three Wars*, 168.

117 Blackwelder, *Long Road*, 15–17. Summarizing data from a number of USAF sources, Blackwelder records consistent weapons effectiveness of greater than 50 percent and CEPs less than 30 feet.

118 From the limited data available, dud and no-guide rates ranged between 19.9–26.1 percent.

119 Gillespie, *Precision Guided Munitions*, 168.

Manner of Employment The next line of inquiry seeks to determine whether PGMs were being employed in a manner commensurate with reducing human suffering. Here, the hybrid nature of this war complicates, as it is clear that the Johnson Administration, and its military subordinates, were taking deliberate steps to minimize civilian casualties long before PGMs became operational in theater. In other words, in sharp contrast to the two previous wars the U.S. had engaged in, where civilians were at times deliberately targeted, national policy explicitly prohibited this practice in Vietnam. As Ward Thomas observes, "Here the commitment to non-combatant immunity went beyond the nominal restriction of raids to military targets ... Targets in populated areas were to be generally avoided, and surface-to-air missile (SAM) sites located in populated areas could only be hit if they were in the act of firing at U.S. planes."[120]

An early noteworthy example can be found in high-level exchanges during the run-up to the June–July 1966 strikes on North Vietnamese POL facilities. In a communication to Admiral Sharp (CINCPAC), informing him of the discussions on the POL attacks, McNamara stated: "Final decision for or against will be influenced by extent they can be carried out without significant civilian casualties. What preliminary steps to minimize would you recommend and if taken what number of casualties do you believe would result?"[121] Admiral Sharp replied that;

> He would execute only under favorable weather conditions, with good visibility and no cloud cover, in order to assure positive identification of the targets and improved strike accuracy; select the best axis of attack to avoid populated areas; select weapons with optimum ballistic characteristics for precision; make maximum use of ECM support in order to hamper SA-2 and AAA radars and reduce "pilot distraction" during the strikes; and employ the most experienced pilots, thoroughly briefed.[122]

The final execution message reiterated these measures to the forces and directed that the mission be aborted if they could not be adhered to.

Another affirmation of the Johnson Administration's active measures to avoid unnecessary civilian casualties can be seen in an exchange between General McConnell and Senator Jack Miller during sub-committee testimony given in 1967. In the context of a discussion about the recently expanded authority to strike targets in North Vietnam, Senator Miller asks, "Are there still some targets

120 Ward Thomas, *Ethics of Destruction: Norms and Force in International Relations* (Ithaca, NY: Cornell University Press, 2001), 152.

121 Department of Defense, *Pentagon Papers*, vol. 6, sec. IV.C.7.(a), 124.

122 Ibid., 124–5; Littauer and Uphoff, eds., *Air War in Indochina*, 39–40, 40n.31; Thomas, *Ethics of Destruction*, 153.

that you have recommended but which have not been authorized because of the civilian casualty consideration?" to which General McConnell responds "Yes."[123]

The introduction of more modern PGMs into the inventory enabled targets to be struck that were much closer to non-military structures than ever before, and the addition of this capability factored into targeting decisions. During Operation LINEBACKER I, for example, concern over civilian casualties clearly influenced weapon selection. President Nixon expressly forbade attacks expected to have high civilian casualties, while the "original LINEBACKER directive stated: 'It is essential that strike forces exercise care in weapons selection to minimize civilian casualties ... [and avoid hitting] known or suspected PW [Prisoner of War] camps, hospitals, and religious shrines.'"[124] Such provisions had certainly existed before; however, the availability of PGMs both improved compliance and allowed for more key targets to be engaged.

Such considerations appear to have been internalized by the planners because "[a]gainst area targets such as railroad yards and storage facilities, where the risk of civilian casualties was minimal, fighters dropped conventional 'iron' bombs." PGMs, however, were used on more difficult precision targets because they used laser or electro-optical guidance and "could hit targets in heavily populated areas with remarkable accuracy."[125] In one notable case, "using guided ordnance, a flight of F-4s destroyed the generator of the Lang Chi Hydroelectric Plant, leaving the dam 50 feet away untouched."[126]

It is evident that, driven from the top by both Johnson and Nixon, there was an explicit desire to plan, conduct, and indeed limit if necessary, air operations in such a manner so as to reduce unnecessary human suffering, generally articulated during that period as "civilian casualties." Cynics might claim that such restrictions came more from a desire to minimize negative public opinion than any true humanitarian desire to lessen the ravages that war visits upon the populations involved. But that is actually a moot point. What matters is that a deliberate effort was made to reduce human suffering, an effort enabled by the introduction of PGMs. And, in North Vietnam at least, this desire seems to have been realized.

Without question, Operation ROLLING THUNDER had a significant impact on the domestic economy of North Vietnam, and caused considerable suffering for its civilian population. National Security Study Memorandum 1 (NSSM 1) reports that:

> Individual civilians suffered many hardships. While the total supply of goods in NVN increased, individual standards of living declined. Food was rationed and consumer goods were scarce; and air raid warnings disrupted the lives of the

---

123 Preparedness Investigating Subcommittee, *Air War Against North Vietnam*, 90th Cong., 1st sess., part 3, 22–23 August 1967, 219.

124 Quoted in Clodfelter, *Limits of Airpower*, 164.

125 Ibid., 159; Thomas, *Ethics of Destruction*, 153.

126 Ibid., 164.

populace and forced many to leave their homes. Moreover, it has been estimated that approximately 52,000 civilians were killed in NVN by U.S. air strikes.[127]

Interestingly, when McNamara was asked, in the summer of 1967, whether American bombing in the North had had an impact on the civilian standard of living, he replied "there has been some hardship on the people, but not a noticeable one."[128]

It is readily agreed that the estimated number of North Vietnamese civilian deaths is not a trivial number. But to place that number in its historical context it needs to be compared with the impact that bombing has had on civilian populations in previous wars. This number, albeit large as an absolute value, represented but 0.3 percent of the population of North Vietnam at the commencement of ROLLING THUNDER. It is significantly smaller, in sheer numbers and in terms of the percentage of the population affected, when compared to Japan (2.2 million/3.0 percent) and Germany (1.1 million/1.6 percent) in World War II, this despite a much greater magnitude and longer duration.[129]

Even staunch anti-war writers admitted the death and destruction was far less than reporting from the North Vietnamese press would indicate. "James Cameron, an antiwar British journalist, toured the North and admitted: 'It was my impression ... that United States attacks on North Vietnam had been, as they claimed, aimed generally at what they could define as military objectives. ... I do not believe that the Americans set out to bomb homes and hospitals; if they had wished to do so it would have been extremely easy to do so.'"[130]

During the LINEBACKER II campaign late in the war, "Acknowledging the raids' accuracy, the journalist Tammy Arbuckle observed during a trip to Hanoi in March 1973: 'Pictures and some press reports had given the visitor the impression Hanoi had suffered badly in the war—but in fact the city is hardly touched."[131] While the data is necessarily incomplete, and the issue remains open for debate, it can be reasonably concluded that this lack of destruction, and relatively few civilian casualties (Hanoi 1,318 killed and 1,216 wounded/ Haiphong 305 dead), is at least partially attributable to the expanded use of PGMs at this later stage of the war.

---

127 Quoted in Committee on Foreign Relations, *Bombing as a Policy Tool in Vietnam: Effectiveness*, 14. Original in *Congressional Record*, Vol. 118, No. 77 (May 11, 1972), p. E5063; James Webb, in the forward to Bui Tin's *From Enemy to Friend: A North Vietnamese Perspective on the War* (Annapolis, MD: Naval Institute Press, 2002), quotes an internal North Vietnamese Joint General Staff study which lists the civilian casualty figure as 79,443.

128 Preparedness Investigating Subcommittee, *Air War Against North Vietnam*, 90th Cong., 1st sess., part 4, 25 August 1967, 341.

129 Pape, *Bombing to Win*, 190.

130 Quoted in Thomas, *Ethics of Destruction*, 154.

131 Clodfelter, *Limits of Airpower*, 195.

Against this second line of inquiry's two metrics then, the trend line can be seen as at least partially positive. There is ample evidence of policy guidance and deliberate planning to reduce suffering and evidence of mandatory restrictions to facilitate that goal. What is difficult to reconcile against this positive determination is the sheer magnitude of human suffering experienced elsewhere in Indochina, as outlined in the earlier narrative.

The government's laudable desire to reduce human suffering, so evident in the literature, was ultimately limited by the technological means available to prosecute the war. In the days prior to the introduction of PGMs into the inventory, the military, inevitably, had to rely on the equipment they had on hand to accomplish their mission. This meant the use of far less accurate tactical aircraft, area coverage weapons, and the large area bombing capability of B-52s. Here, perhaps, is one instance where the Vietnam War case study serves well as a baseline for future cases to be measured against.

Part II: The Question of Unintended Consequences

The Vietnam case study may seem, at first glance, to offer little substantive evidence for this portion of the analytical framework. With PGMs comprising less than one percent of the overall ordnance dropped in all of Southeast Asia during the war, it would seem nearly impossible to be able to differentiate between any unintended consequences, due to aerial bombing, caused by conventional or from "smart" bombs.[132] However, a careful review of the evidence reveals that there are in fact lessons to be drawn that will be useful in the wider analysis.

Rapid capitulation with unresolved tensions With the expanded war underway for more than two years before the deployment of the first PGMs to the Indochina, coupled with the fact that there never was a "rapid capitulation" by the North Vietnamese, this line of inquiry at first seems fruitless. But there is an important insight to be gained into the mindset of the senior American policy makers during this period, which may inform the analysis in later case studies. As Kipp's quote in the epigraph alludes, the war in Vietnam stands near the beginning of a conscious, conceptual transformation in the minds of senior civilian and military leaders regarding the use of American military power.

Following on from two wars where air power played a key, but mainly supporting role in what were primarily massive land and amphibious campaigns, a paradigm shift seemed underway. Now, with technology as an enabler, carefully focused firepower, delivered from the air, was increasingly seen by many as replacing manpower as the primary method of bending the enemy's will to meet American foreign policy objectives.[133]

132 Blackwelder, *Long Road*, 18–19.

133 Again, many eventually will come to call this technology driven effort the RMA.

While not all agreed, General Harold K. Jones key among them, it is evident from the early debates at the highest levels that many were convinced that a focused application of American air power against North Vietnam would convince Hanoi, with relatively little bloodshed or need to deploy American ground forces, to cease supporting the insurgency in the South with both men and equipment. In other words, it was believed that the proper application of air power *could*, even *would*, result in a rapid capitulation by the enemy.[134]

Significantly, just as the desire to limit civilian casualties predated the introduction of PGMs, this fundamental belief in the coercive potential of air power was held *before* we actually had the technological means to apply force as precisely as was required, although clearly some believed that the technology was mature enough by 1967. This professed ability to win quickly and cheaply, anathema to the Prussian theorist Clausewitz, comes close to Sun Tzu's dictum that it is best to win without fighting.[135] Clearly, if attainable, this would be a seductive course of action for decision-makers.

Adverse asymmetrical responses Once again, it cannot logically be argued that an American reliance on precision-guided munitions forced this particular enemy into an asymmetrical response. The war was already underway, and the North Vietnamese and their proxies had already adopted insurgent tactics, well before PGMs were used. That strategy however, patterned on Mao's three-phase model, was at least in part based on a determination by the leadership in North Vietnam that it could not go head-to-head with the U.S. military, viewed at the time as the best in the world.[136] However, the outcome of this apparent mismatch between the insurgent tactics used by the North and the more technological methods by which the Americans sought to counter them is very enlightening. What is also illuminating is what happened to North Vietnamese forces when they abandoned those insurgent tactics later in the war. Combined, these two items offer a good deal of baseline insight.

When examining the first, it is important to recognize that guerrilla forces have a number of comparative advantages when facing conventional opponents. Among these: the ability to draw strength, in terms of support, food, and recruits, from the

134 The possibilities of such coercion from the air had long been touted by airpower theorists such as the Italian Giulio Douhet.

135 Clausewitz belittles those who think that victory can be "achieved without bloodshed," while Sun Tzu believes that "winning without fighting" is the acme of skill. See Clausewitz, *On War*, 75–6 and Sun, *Art of War*, 79.

136 The North Vietnamese initially viewed the American fighting man, ten feet tall and nearly invincible. See Bui Tin, *From Enemy to Friend*, 13–14. He points out that General Vo Nguyen Giap was very concerned by the possibility of a U.S. ground intervention in 1965. Mao's three-phased approach is outlined in the essay "On Protracted War." See Mao Tse-Tung, *Selected Military Writing's of Mao Tse-Tung* (Peking (Beijing): Foreign Languages Press, 1967), 210–19.

populace; the ability to blend into the population when necessary to elude opposing forces; and the ability to use their mobility, and generally smaller size, to pick the time and place of an engagement.[137] These attributes mean that it is difficult for a conventionally-structured force, particularly one that is overly reliant on advanced technology, to localize, target, and destroy guerrilla forces in a traditional force-on-force fashion. There simply are, in many cases, no conventional enemy forces to engage. In such cases the population itself becomes the principal strategic target. Killing innocent civilians is obviously counterproductive and unethical, but, paradoxically, "killing off the guerrillas does not necessarily deplete their number. In the French-Algerian war of 1954–1962, for example, in order to reduce the rebel forces from 40,000 to 12,000 men, the French army had to kill 120,000 insurgents and capture another 60,000."[138]

What all of this indicates is that the preferred American "way of warfare" at that time, heavily reliant on advanced technology, a high degree of mechanization, and air power, which was well-suited to the nation's main strategic concerns along the Central Front in Europe, was ill-suited for the war the North Vietnamese fought. As one expert on insurgency, Roger Hilsman, noted, "the best weapon for fighting the guerrilla is a knife, the worst is a bomber; the second best is a rifle, the second worst, artillery."[139] With the adoption of an even greater technological focus today, to include a steadily increasing reliance on PGMs, an opponent who adopts an asymmetric approach, insurgency or otherwise, would seem to continue to cause the American military much concern.[140]

In fact, in a 1975 Rand study examining the possible implications of this new method of LGB-enabled, precision-guided warfare, and how such warfare might look in 20 years, James Digby offered a number of propositions, among them an early recognition that "concealment will become more important" as a target seen by the targeting aircraft is now highly likely to be destroyed.[141] Concealment can, of course, take many forms ranging from traditional camouflage, to simply blending into the society at large.

The second insight emerges out of the wreckage of the North Vietnamese Easter Offensive of 1972. Shifting to Mao's third phase, a more traditional conventional force-on-force strategy, the North Vietnamese played directly into the strengths of the American forces. Out of hiding, and much more concentrated, their forces were much easier prey for American aircraft, now armed with PGMs. In this Clausewitzian world of interaction and adaptation, the lessons for both those who would oppose the U.S., and for those who must order the use of lethal force on her behalf, appear straightforward. For the former group an early example that,

137 For an overview see Littauer and Uphoff, eds., *Air War in Indochina*, 3.

138 Littauer and Uphoff, eds., *Air War in Indochina*, 3.

139 Quoted in Littauer and Uphoff, eds., *Air War in Indochina*, 51.

140 Perhaps as seen in Iraq after the fall of the regime in 2003.

141 James F. Digby, *Precision-Guided Weapons: New Chances to Deal with Old Dangers*, Rand no. 5384 (Santa Monica, CA, 1975), 8.

unless you are a peer competitor, playing to American strengths is not likely to be a war-winning strategy. For the latter, a reminder of the need to remember that an adaptive adversary may fight asymmetrically, and that planners need to adequately address that possibility before operations commence.

Long term suffering This line of inquiry examines whether the use of PGMs, while initially resulting in a lower level of human suffering due to their accuracy and judicious employment, actually result in a greater degree of human suffering due to long-term damage caused to critical infrastructure. While the hybrid nature of the conflict creates some difficulties in this analysis as well, once again there is evidence that the American government was carefully considering such issues, before PGMs were introduced into the theater. This war also highlights another intervening trend which can serve to complicate the analysis.

If, as Arthur Ponsonby posited, "When war is declared, Truth is the first casualty," then surely it takes years to cull through the accumulated deceit to regain it.[142] Throughout the war, the government in Hanoi mounted an effective propaganda counter-offensive against the American bombing campaign, alleging a great degree of human suffering. It, along with similar anti-war efforts, were so successful in blurring fact from fiction, that Gunter Lewy felt compelled to write *America in Vietnam* in an effort to bring "to bear more light, rather than heat, on the discussion of American involvement in Vietnam."[143] One example of such propaganda campaigns, and one that Pape uses to illustrate this point, is that of the dike system in North Vietnam.

While it is a historical fact that dikes were once seen as legitimate targets, and had, in fact, been attacked during both World War II and the Korean War, this was not the case in Vietnam. The North Vietnamese did charge "the United States with the devastation of the dike system, the bulwark of North Vietnam's agricultural economy. [If true], the destruction of dikes could flood rice paddies and so threaten a basic staple of the civilian diet, and floods could threaten civilians directly."[144] Such an action would likely have caused significant long term suffering for the North Vietnamese people. Hanoi's charges, however, were false.

During the war, the American government never systematically targeted the 91 dike system, although a small number of them were in fact hit. In 1967, a claim was made that the American's had breached 24 of them. When these claims were investigated, "U.S. intelligence determined that only four had been partially damaged by accident, with no evidence of flooding."[145] In actual practice, the

142 This quote is often attributed to Senator Hiram Johnson purportedly while addressing the U.S. Senate in 1918. Others, however, point out that it is not be found in the record of the speech. It is contained in Ponsonby's *Falsehood in Wartime* (1928). Yet others credit Samuel Johnson (1758) or the Greek writer/poet Aeschylus.

143 Lewy, *America in Vietnam*, v.

144 Pape, *Bombing to Win*, 189–90.

145 Ibid., 190.

North Vietnamese clearly recognized this restriction and used the dikes as staging areas for anti-aircraft equipment, precisely because they knew that they would not be attacked.[146] Even the eventual change in administration did not remove this restriction. During LINEBACKER, President Nixon, too, "forbade the bombing of dams 'because the results in terms of civilian casualties would be extraordinary.'"[147]

Two other observations also help demonstrate that the level of human suffering, both short and long term, was probably lower than that portrayed at the time. First, in May of 1967, U.S. air strikes specifically targeted and successfully destroyed 14 of 20 electrical power facilities in North Vietnam, reducing national capacity by 87 percent. Yet, this coordinated attack on one of the few "modern" sectors of North Vietnam's economy, which elicited great claims of civilian casualties from Hanoi, did not achieve its goal. Although "highly successful in pure target-destruction terms, [it] had failed to alter Hanoi's determined pursuit of war."[148] More importantly, portable generators quickly replaced the lost power through much of the city.

By the fall of 1967 aerial operations had exacted a significant toll on the North Vietnamese economy and population. This might reasonably be expected to lead to a high level of human suffering. Yet a December 1967 report, known as the JASON Study, concluded:

> Measurable physical damage now exceeds $370 million and the regime has had to divert 300,000 to 600,000 people (many on a part-time basis) from agricultural and other tasks to counter the bombing and cope with its effects. The former cost has been more than met by aid from other Communist countries. The latter may not be real, since the extra manpower needs have largely been met from what was a considerable amount of slack in NVN's under-employed agricultural labor force.[149]

The report goes on to explain that North Vietnam had enough manpower to operate the economy, send troops to the South, and to defend the homeland.

The preceding examples serve to make two points. The first is that in this conflict, even before PGMs were available, senior leaders were making conscious decisions designed to avoid/reduce long term suffering to civilians in North Vietnam. The second is that determining the effect of bombing on civilians, particularly during the conflict, is a difficult task. Actions which would at first appear to have the potential for creating long term suffering, and which may elicit claims of such, may in fact not be true because the claims are deliberately falsified or exaggerated, or as societies find "work around" solutions.

146 Thomas, *Ethics of Destruction*, 153.

147 Clodfelter, *Limits of Airpower*, 164.

148 Department of Defense, *Pentagon Papers*, vol. 6, sec. IV.C.7.(b), 20.

149 Ibid., 20, 128–9.

Two additional observations also emerge. As noted previously, human suffering in the other sub-theaters of the war was significant. For example, "Between 1964 and 1969, as many as 3.5 million South Vietnamese, over 20 percent of the population, had been refugees at one time or another. This figure did not include people only temporarily displaced by acts of war, such as the almost 1 million persons whose houses were destroyed during the enemy's Tet and post-Tet offensive in 1968."[150] In these areas, the heavy use of area coverage weapons and B-52 area bombing combined to create higher levels of devastation and long term suffering. This helps to establish a baseline for comparison in later case studies, as the evidence points to less widespread devastation when PGMs were used.

Finally, the issue of unexploded, or "dud," ordnance must be acknowledged. Here the issue of long term suffering is directly tied to the inadvertent detonation of a long dormant bomb or sub-munition, at times by unsuspecting farmers, at others by children playing with a new found "toy," long after the conflict is over. This is not dissimilar to what happens with land-mines, and is subject to much the same criticism. While Lewy makes a convincing argument that CBUs, for example, are legal, for this study's purposes that misses the larger point. CBUs do have high dud rates, as the Cornell study notes, and they do present an ongoing threat to the local population. One positive aspect of PGMs is that far fewer are required to accomplish the mission, and they have a relatively high rate of operational effectiveness, thus lowering the potential for future suffering.[151]

Guenter Lewy, in *America in Vietnam*, concludes that, despite the popular mythology that emerged from this period, "detailed examination of battlefield practices reveals that the loss of civilian life in Vietnam was less great than in World War II and Korea and that *concern with minimizing the ravages of war was strong*."[152] While acknowledging the concerns of those who find any resort to war repugnant, he believes that "as long as wars do take place at all it remains a moral duty to seek to reduce the agony caused by war, and the fulfillment of this obligation should not be disdained."[153] It is the role of PGMs in this latter duty that is being examined.

Part III: The Question of Teleology

Unfortunately, appearing repetitive, it is once again difficult, at first glance, to see how this third level of analysis can be conducted as the availability of PGMs, actually the non-availability in this case, could not have influenced key decision makers during the run up to offensive combat operations in Southeast Asia. However, what does emerge from the literature dealing with discussions at the highest levels of the American government is interesting evidence regarding the

---

150 Lewy, *America in Vietnam*, 108.
151 Ibid., 267; Littauer and Uphoff, eds., *Air War in Indochina*, 56.
152 Ibid., vii: emphasis mine.
153 Ibid.

broader considerations regarding the use of air power. Today, with strike warfare from the air domain enhanced and enabled by the ready availability of a wide range of PGMs, it will be interesting to determine if the early considerations evidenced here have matured in contemporary times.

Reluctance to deploy conventional land forces The ability to use offensive airpower as a means to attempt the coercion of the government of North Vietnam allowed America to initially limit its use of offensive ground troops to South Vietnam, although eventually forces spread to Cambodia and Laos. This limitation was very understandable given Washington's fear of direct intervention by Hanoi's two patron states, the People's Republic of China and the Soviet Union, as the former had done in Korea. While it is unlikely that there was greater human suffering in North Vietnam due to the reluctance of the American government to introduce ground forces, as might be the case in a nation the country was seeking to liberate or overthrow, the effect on South Vietnam is less clear.

While it might be convenient to argue that a sufficient number of troops, deployed early in the war to portions of North Vietnam, would have stemmed the flow of troops and supplies to the South thereby shortening the war and reducing human suffering, it is simply unsupportable. No reliable evidence exists that would allow for the creation of a convincing scenario that accounted for all the possible "chess moves" between at least five key participants.[154] But what is clear is that, for the first time, the U.S. exhibited a willingness to use air power alone, as a potentially war-winning tool. This tendency was also observed by the Cornell Group, albeit in a different context. It noted, "In this sense, the Laos campaign has become a clear application of that aspect of the Nixon Doctrine which rests upon the substitution of American technology, in conjunction with Asian forces, for American manpower."[155]

Greater willingness to resort to lethal force As the war in Southeast Asia was underway before such weapons were available, it would be specious to argue that the Johnson and Nixon Administrations exhibited a greater willingness to resort to lethal force due to the presence of PGMs. What is clear, however, is that senior civilian and military leaders were increasingly willing to authorize strikes into more heavily populated areas as the enhanced capabilities of these weapons was realized.

Littauer and Uphoff record that in 1971, "smart bombs reportedly accounted for less than one percent of the total weight of munitions dropped in Indochina," a statistic that actually holds true for the entire war if only tactical aircraft sorties are analyzed.[156] An Institute of Defense Analysis study done in 1974, "Evaluation of

154 The United States and South Vietnam on one side, the Soviet Union, the People's Republic of China, and North Vietnam on the other.

155 Littauer and Uphoff, eds., *Air War in Indochina*, 80.

156 Littauer and Uphoff, eds., *Air War in Indochina*, 156n.23.

Guided Bomb Systems Employed in SEA," determined that during the war 26,690 PGMs were dropped, which was less than one percent of the estimated 3,376,000 "dumb" bombs dropped during the same period.[157] But this "one-percent" has had significant long term implications.

As Gillespie observes, "as America emerged from the Vietnam conflict, military and government officials were convinced that 'surgical' air strikes were not only possible, but were, in fact, routine. The following statement from the Air Force Secretary's June 1972 'Policy Letter for Commanders' typifies the official reaction to this newly demonstrated capability:

> New weapons and tactics resulting from the accelerated air war have significance not only in this war but in evaluations of strategy to be used elsewhere in the future. The unprecedented accuracy of laser-guided or TV-guided 'smart' bombs and airborne sensors now being used by U.S. aircraft is making interdiction far more effective than before."[158]

Simply put, PGMs were noticed.

Skolnikoff sees this as a component part of a wider trend. "As a result of the ineffectiveness of some high-technology weapons against determined guerilla forces that was shown in Vietnam, increased [research and development] attention was directed to applying technology to the conventional battlefield. The 'electronic battlefield' of sensors, antipersonnel weapons, and precision-guided 'smart' munitions followed."[159] Here, perhaps, is the beginning of the RMA.

The Vietnam War, a transitional case study in terms of the availability of PGMs, affords some solid conclusions at one level of the evaluative framework, and some interesting insights at the other two. First, it is clear that even at this early juncture in their employment, PGMs do, in fact, exhibit both greater accuracy and a high degree of reliability. It is also evident that decision makers from the highest levels on down are concerned with both how and when they will be employed, with an eye towards lowering civilian casualties and collateral damage. Interestingly, these latter concerns pre-date the arrival of these new weapons in theater.

It is also apparent, that civilian and military leaders were concerned with the potential for long term suffering before 1967, to such a degree that it deliberately placed aircrew at greater risk to avoid catastrophe. What will be interesting to determine, as additional cases are examined, is whether such concerns persist over time and to what degree the expanded use of PGMs does, or does not, affect these dynamics.

---

157 Quoted in Blackwelder, *Long Road*, 18–19. He further notes that this includes bombs delivered by both the U.S. Air Force and U.S. Navy aircrews. Of this total, 94 percent were LGBs, the remaining 6 percent were EO-guided weapons.

158 Gillespie, *Precision Guided Munitions*, 175–5.

159 Skolnikoff, *Elusive Transformation*, 72.

Finally, while inconclusive at this point, early evidence would suggest that, while the U.S. appears to be willing to use air power alone, and perhaps more so after PGMs are demonstrated to be highly effective, there is no clear evidence that leaders are willing to use violence more frequently. Only further analysis will answer that question.

Chapter 5
Desert Storm: An Image of Perfect Precision?

There goes the luckiest man in Baghdad.

General H. Norman Schwarzkopf

Because of advances in media technology, the use of PGMs in the 1990s garnered international attention and high popular expectations. While technique retains its position of primacy, another of Jacques Ellul's principal concerns with modern society is humankind's strong addiction to the "image," and the accompanying diminution, he calls it the humiliation, of the importance of the spoken word through which a deeper understanding of issues could be obtained.[1] The epigraph above is perhaps a telling example. For those who have seen it, who can forget the image portrayed on that short video clip? Through the "eye" of the targeting system of a coalition aircraft, a private vehicle is seen crossing over a bridge near Baghdad in early 1991. Literally a second or so after the car goes past, the span is destroyed by an LGB dropped from above.

Narrated by General H. Norman Schwarzkopf, Commander-in-Chief of the U.S. Central Command and the senior U.S. military officer in-theater, at one of his daily press briefings in Riyadh, the scene portrayed sent a clear message; "look how careful we are being to ensure that non-combatants aren't hurt." But such images, played repeatedly before a worldwide television audience, also helped implant in the minds of the viewers the idea that this was a sanitized, "push-button" precision war. The actual story of the Gulf air war, and the role PGMs played in it, is much more complicated. Of all the munitions dropped during Operation DESERT STORM, only about 7.5 percent were actually PGMs, the rest were conventional, unguided munitions, in other words, "dumb bombs."[2]

That the nearly perfectly flat, open terrain of the theater of operations favored the American technological approach to warfare is almost universally acknowledged.[3] It is also true that Coalition planners, as was witnessed in the planning of air operations into North Vietnam, went to great lengths to ensure that precision weapons, often

1 This concept will be returned to later.

2 The definitive post-war study records that of some 225,000 bombs dropped during the war, 17,000 were PGMs. This statistic can be misleading depending on which munitions are included. For consistency, this study will only examine the PGMs and cruise missiles. Thomas A. Keaney and Eliot A. Cohen, *Gulf War Air Power Survey*, Summary Report, (Washington, DC: U.S. Government Printing Office, 1993), 226. Hereafter *GWAPS*.

3 Skolnikoff and Van Creveld among them.

dropped by new stealthy aircraft, were used in those areas where the risk of collateral damage, and increased human suffering, were greatest.

But in other places, particularly against entrenched troops in the featureless desert along the front lines, and against large area targets such as airfields, planners once again returned to familiar weapons and delivery techniques. Using mass, or dispersion, to replace precision where feasible, or required by the lack of available PGM-capable aircraft, even stalwart Vietnam workhorses like the B-52 played important roles in this war.[4]

The Decade before the Storm

To understand the political, social, and economic realities that convinced Saddam Hussein that Iraq needed to invade its neighbor Kuwait in August 1990, and why an unprecedented coalition of Arab and Western militaries formed to oppose it, it is necessary to have a broad understanding of the events of the preceding decade. For most of that period, Iraq was engaged in a bloody regional conflict with Iran. During that war, a war which it had initiated, Iraq greatly expanded its already sizeable army and simultaneously incurred a massive foreign debt. Waged between 22 September 1980 and 20 August 1988, the early part of the war saw a number of Iraqi successes. The conflict, however, quickly turned into a virtual stalemate and from about mid-1982 onwards Iraq found itself largely on the defensive.[5] While Saddam was largely able to convince the people of Iraq that the war was a victory for Iraq, the nation had undeniably been crippled by the extended effort.

Although starting the war with about $35 billion in foreign exchanges reserves, almost from the very beginning Iraq was forced to fund its war effort using foreign loans.[6] By war's end, Iraq owed about $50.5 billion, a total which was some three times Iraq's annual Gross Domestic Product (GDP).[7] Even more problematic for Saddam in the long term, external estimates for Iraq's post-war reconstruction ranged as high as $230 billion.[8]

At the same time, Iraq had greatly expanded its military forces, which grew from around 240,000 to 1.2 million personnel.[9] To compensate for this displacement

---

4 This was driven by necessity as most American aircraft in 1990–1991 were unable to deliver PGMs.

5 Dilip Hiro, *Desert Shield to Desert Storm: The Second Gulf War* (London: Paladin, 1992), 44. Hiro argues that Saddam would have preferred the war end in June 1982 when Israel invaded Lebanon.

6 Lawrence Freedman and Efraim Karsh, *The Gulf Conflict 1990–1991: Diplomacy and War in the New World Order* (Princeton, NJ: Princeton University Press, 1993), 39.

7 Ibid., 1, 42–3. Others put the debt at $80 billion. Freedman and Karsh, *Gulf Conflict 1990–1991*, 39; Norman Friedman, *Desert Victory: The War for Kuwait* (Annapolis: Naval Institute Press, 1991), 31.

8 Freedman and Karsh, *Gulf Conflict 1990–1991*, 39.

9 Hiro, *Desert Shield to Desert Storm*, 1.

of manpower, migrant workers had flooded Iraq's civilian workforce. How to reintegrate thousands of demobilizing soldiers into this struggling economy posed a significant problem for Saddam. Dilip Hiro, author of *The Longest War*, an acclaimed history of the Iran-Iraq War, believes "these [two] factors came to play key roles in the two-year interregnum between the end of the First Gulf War and the Iraqi Invasion of Kuwait."[10] Further compounding this already complex economic crisis was the fact that the price of oil, which accounted for roughly 95 percent of Iraq's income and which was its only real hope for fiscal relief, was falling.[11]

To address the second of these issues, that of demobilization, Saddam forced out the roughly two million migrant workers who had alleviated the labor shortage during the war.[12] While this provided employment opportunities for the returning soldiers, and simultaneously lowered the overall costs for maintaining the Iraqi military, even this effort was not enough to save the economy. In fact, Iraq's total annual oil revenues only just covered the annual military budget, leaving the country about $10 billion short of that needed to balance the state budget, even before any money could be spent on the much needed reconstruction.[13] Saddam had, broadly speaking, two potential solutions to his economic crisis. One, he could somehow engineer an increase in the price of oil or, two, he could obtain money from outside sources. He attempted both.

Pursuing the former course involved trying to get the OPEC nations to abide by the quota system first set up by the organization in 1986, a system that was both complex and routinely ignored by member states. In early 1990, with OPEC's production almost 10 percent above the target of 22 million barrels per day, Saddam began to increase pressure.[14] In a February 1990 meeting in Kuwait, the Iraqi Oil Minister Isam Abd al-Rahim al-Chalabi "pressured his hosts to abide by the new oil quota set by OPEC earlier that year. Then he proceeded to Riyadh to deliver a personal message from Saddam Hussein to King Fahd. The Saudi's must convince the rest of the Gulf states not to exceed their oil quotas."[15]

In May, at an OPEC meeting in Geneva, Iraq again pressed hard for an adherence to the quotas by the Gulf oil states, with Saddam himself raising the stakes at this meeting. Pointing out that each one dollar drop in the per barrel price of oil cost Iraq $1 billion per year, he declared that "The continued violation of oil quotes by some Arab states amounted to a *declaration of war on Iraq*."[16]

---

10 Ibid. For an in-depth history of the Iran-Iraq War, see Dilip Hiro, *The Longest War: The Iran-Iraq Conflict* (New York: Routledge, 1991).

11 Friedman, *Desert Victory*, 31; Freedman and Karsh, *Gulf Conflict 1990–1991*, 40; Hiro, *Desert Shield to Desert Storm*, 53.

12 Freedman and Karsh, *Gulf Conflict 1990–1991*, 39.

13 Ibid.

14 Ibid., 41.

15 Ibid., 45.

16 Ibid., 46: emphasis in original.

Simultaneously, on the second front, Saddam returned to a theme he had first employed during the Iran-Iraq War. During that conflict, he had tried to convince his neighbors that the war "was not Iraq's private business … but rather a defence of the eastern flank of the Arab world against fundamentalist Islam. While the Gulf states were not asked to pay with rivers of blood for the protection of their security, since Iraq did that on their behalf, they could not expect to take a 'free ride' on Iraq's heroic struggle."[17]

In Amman, Jordan, at a February 1990 meeting of the Arab Co-operation Council, Saddam enlisted the aid of Egyptian President Hosni Mubarak and Jordan's King Hussein, asking them to let the Gulf states know he "not only [expected] a complete moratorium on its wartime loans, but urgently needed an immediate infusion of additional funds of some $30 billion. [Warning,] 'Let the Gulf regimes know … that if they do not give this money to me, I will know how to get it.'"[18] Once again, Saddam did not receive a response that he considered satisfactory. By July, tensions in the region were very high. Concerned Arab and Western diplomats sought to find a way to ease regional tensions.

The story of the complicated diplomacy that took place during the last three weeks of July 1990 has been ably related elsewhere.[19] In the end, a non-military solution to the crisis could not be found. Despite apparent assurances to Egypt's President Mubarak and the American Ambassador to Iraq, April Glaspie, that he would not invade Kuwait, Saddam did just that. At 0100 on 2 August 1990, Saddam became, in the words of Norman Friedman, "a debtor canceling his debts by robbing the bank."[20]

The Crisis Unfolds

Despite heroic efforts by some individual Kuwaiti units, Kuwaiti Air Force A-4Ku aircraft, for example, fought for three days using a local road as a runway, until their fuel and ammunition were depleted; for all intents and purposes Kuwait was annexed by Iraq within twelve hours.[21] This apparent *fait accompli* presented national leaders in both the Middle East and the West with a difficult situation, brought about not only by the brashness of the act itself, but also by the immediate recognition of the significant amount of oil that Saddam Hussein now controlled.[22]

---

17 Ibid., 45.

18 Ibid., 45.

19 Representative accounts can be found in: Freedman and Karsh, *Gulf Conflict 1990–1991*, 42–63; Hiro, *Desert Shield to Desert Storm*, 83–97. For an inside view from the White House see George [Herbert Walker] Bush and Brent Scowcroft, *The World Transformed* (New York: Vintage Books, 1998), 302–14.

20 Friedman, *Desert Victory*, 31.

21 Ibid., 38; Freedman and Karsh, *Gulf Conflict 1990–1991*, 67.

22 With Kuwait in hand, Saddam controlled "the second- and third-largest proven oil reserves with the fourth-largest army in the world." Bush and Scowcroft, *World*

President George H.W. Bush, heavily conditioned by the events in Europe during the 1930s, believed from the very beginning that the only correct course was to reverse the current state of affairs.

At a meeting with British Prime Minister Margaret Thatcher on 2 August the president relayed the key theme of his conversations with President Mubarak and King Hussein earlier that day: "I said we couldn't accept the status quo. It had to be withdrawal and the restoration of the Kuwaiti government."[23] So began the slow and delicate process of key world leaders, Western and Arab, working with each other and the United Nations to craft a mutually acceptable solution to the problem.[24]

While early discussions between senior U.S. military and civilian leaders focused on any number of key issues, of paramount importance was the requirement to decide just what the nation was willing to fight for. While President Bush focused on the liberation of Kuwait quite quickly, at these early stages in the crisis not all thought that was the correct course of action. At the Pentagon, Secretary of Defense (SECDEF) Cheney, while agreeing with the president's goal, went so far as to wonder if that option was itself too limited and if the overthrow of Saddam should be the objective. But, significantly, General Colin Powell, the Chairman of the Joint Chiefs-of-Staff, was solidly in opposition.[25]

Powell firmly believed that Kuwait was not worth the "blood and treasure" to be expended. At a 2 August meeting with Cheney and others, he declared, "The American people do not want their young dying for $1.50 gallon oil ... We can't make a case for losing lives for Kuwait, but Saudi Arabia is different. I am opposed to dramatic action without the President having popular support."[26] While Powell continued to have concerns with regards to Kuwait, within two days a solidifying of positions within the Administration and beyond seemed underway.

Speaking to the press on 5 August President Bush publicly referred for the first time to the liberation of Kuwait, stating:

> What's emerging is nobody seems to be showing up as willing to accept anything less than total withdrawal from Kuwait of the Iraqi forces, and no puppet regime. We've been down that road, and there will be no puppet regime that will be accepted by any countries that I'm familiar with. And there seems to be a

---

Transformed, 322.

23 Bush and Scowcroft, *World Transformed*, 319.

24 Just after the invasion, King Hussein of Jordan and Egyptian President Hosni Mubarak asked President Bush to give them two days to work on an Arab solution. Freedman and Karsh, *Gulf Conflict 1990–1991*, 69; Bush and Scowcroft, *World Transformed*, 318–19.

25 Michael R. Gordon and General Bernard E. Trainor, *The General's War: The Inside Story of the Conflict in the Gulf* (New York: Little, Brown and Company, 1995), 33–34.

26 Ibid.

> united front out there that says Iraq, having committed brutal, naked aggression, ought to get out, and that this concept of their installing some puppet—leaving behind—will not be acceptable … *This will not stand. This will not stand, this aggression against Kuwait.*[27]

Three days later, in an address to the American people announcing the deployment of the first U.S. troops to Saudi Arabia, the president outlined what he called "four simple principles" guiding American policy, principles which were to remain remarkably consistent throughout the crisis and eventual war.

> First, we seek the immediate, unconditional, and complete withdrawal of all Iraqi forces from Kuwait. Second, Kuwait's legitimate government must be restored to replace the puppet regime. And third, my administration, as has been the case with every President from President Roosevelt to President Reagan, is committed to the security and stability of the Persian Gulf. And fourth, I am determined to protect the lives of American citizens abroad.[28]

As would be expected, while such concerted public and private diplomatic and economic efforts were underway, military planning was taking place behind the scenes.[29]

These momentous world events had thrust a relatively unknown, and indeed somewhat marginalized, major U.S. military command, the U.S. Central Command (CENTCOM) into the limelight. In 1990, CENTCOM was not regarded highly by professional military officers.[30] The latest manifestation of an organization that had started out as the Rapid Deployment Force (RDF), a Carter Administration initiative primarily designed to counter any Soviet military moves in the region, CENTCOM had been under the command of Army General H. Norman Schwarzkopf since 23 November 1988.

One of Schwarzkopf's first initiatives upon assuming command was to conduct a thorough review of CENTCOM's primary warfighting focus, a task that was particularly important in light of the declining influence of the Soviet

27 George H.W. Bush, President, Press Conference (White House, 5 September 1990): emphasis mine; Bush and Scowcroft, *World Transformed*, 319, 326, 329.

28 George H.W. Bush, President, Address (White House, 8 August 1990). The first two principles form the main goals for the U.N.-sanctioned Coalition efforts. While the last two were initially U.S. national interests, the third came to mean "stripping Saddam of the means to threaten his neighbors." Pape, *Bombing to Win*, 219. These four objectives were repeated, almost verbatim, in National Security Directive (NSD) 54 "Responding to Iraqi Aggression in the Gulf" issued by the president on the eve of Operation DESERT STORM.

29 President Bush received his first briefing on military options the day after the invasion. Bush and Scowcroft, *World Transformed*, 316.

30 Gordon and Trainor, *General's War*, 42–4. General Schwarzkopf admits as much in his memoir, but his family's history in the region made it an attractive posting for him. H. Norman Schwarzkopf and Peter Petre, *It Doesn't Take a Hero* (New York: Bantam, 1992), 271.

Union in the command's assigned area of responsibility. The initial effort was aided by SECDEF Dick Cheney's first Defense Planning Guidance, a document which provides "firm guidance in the form of goals, priorities, and objectives, including fiscal constraints" to senior military commanders.[31] This classified document's reference to a "robust regional threat" in the Gulf region gave Schwarzkopf the leeway he needed to shift CENTCOM's planning focus from that of a Soviet invasion in Iran, to an Iraqi invasion of the oil fields of Saudi Arabia.[32]

While this shift in the command's strategic emphasis was not without some controversy, some at the State Department expressed serious reservations, Iraq eventually featured as the enemy in CENTCOM's next major biennial exercise INTERNAL LOOK. In an ironic twist of history, INTERNAL LOOK 90 was underway when Saddam invaded Kuwait. But the invasion occurred well before all the plan's myriad details had been completely worked out.[33]

While during the first few days of the crisis the longer-term political objectives seemed to be solidifying, the more immediate tasks at hand were to decide what the U.S. could do in the near-term to dissuade Iraq from continuing its invasion into Saudi Arabia, and what it could do to defend the peninsula if that attack materialized. From the start, it was evident that the sheer distances involved created major problems.[34]

Early estimates from CENTCOM, briefed to the National Security Council at Camp David on 4 August, indicated that it would take roughly a month to get a credible defensive ground force in place, *if* Saudi Arabia permitted the basing of American troops in their country.[35] Until that time, any American response would have to rely on naval forces in, or flowing towards, the theater, and any U.S. Air Forces assets that could quickly be based in the region, or flown the long distances to the battlefield, such as B-52s. Once regional basing was approved by the host nations, Schwarzkopf and his Air Component Commander, U.S. Air Force Lieutenant General Charles Horner, felt they could have sizeable air assets in place in about three weeks and wage a primarily air-centric defensive campaign against

31 Joint Publication 1-02, s.v. Defense Planning Guidance, 151.

32 Schwarzkopf and Petre, *Hero*, 286–8.

33 Gordon and Trainor, *General's War*, 44–6. Schwarzkopf personally did not believe that Saddam would attack, and stated so in a Pentagon briefing on 31 July. Colonel Richard T. Reynolds, *Heart of the Storm: The Genesis of the Air Campaign Against Iraq* (Maxwell AFB, AL: Air University Press, 1995), 1. In his memoir, Schwarzkopf states he thought they would cross the border, but not attempt to take the whole country. Schwarzkopf and Petre, *Hero*, 294–5.

34 Authors, in a number of disciplines, talk of the "tyranny of distance," a concept which applies here.

35 Schwarzkopf eventually felt he had this capability in place on 24 Sept 1990. Gordon and Trainor, *General's War*, 57.

Iraqi forces in about 17 weeks.[36] As for mounting a ground campaign to remove Iraqi forces from Kuwait, the CINC felt that it would take between eight and ten months to deploy and prepare forces for that mission.[37]

At this point, the eventual manpower requirements were truly unknowable. CENTCOM's nascent operational plan, OPLAN 1002-90 "Defense of the Arabian Peninsula," had been in its second draft when Iraq attacked. Consequently, the complicated integrated deployment plan had not yet been completed.[38] Even more troubling, the OPLAN's envisioned timeline had already been blown, and such simple planning assumptions as host nation support had not been realized. With this skeleton in hand, CENTCOM began to plan for the potential coming conflict.

Following Saudi approval, the first American forces arrived in the Kingdom on 8 August 1990, marking the beginning of Operation DESERT SHIELD, the massive deployment of forces and equipment that would continue nearly unabated over the next six months. What was initially briefed to Saudi Ambassador Prince Bandar as a U.S. ground force of some 100,000 to 200,000 troops, grew first to 250,000, then more than doubled to 541,000 prior to the start of the ground war.[39] With the addition of the Coalition partners, almost 800,000 troops eventually waged the campaign to liberate Kuwait from the Iraqis. What finally transpired was a devastating, nearly six-week-long, air campaign, followed by an impressive one hundred-hour ground war. In the end, to the world's grateful astonishment, Iraq was quickly expelled from Kuwait and much of its military was destroyed.

But none of this could be known in August 1990. While many worked on deployment orders and the detailed planning for the potential ground campaign, others, particularly outside of CENTCOM's staff, were working to see how the air forces flowing into theater, with the precision-guided munitions they might employ, could best be brought to bear. It was these latter efforts that eventually proved decisive in DESERT STORM's air campaign.

36 By 17 August, at an INSTANT THUNDER brief given to the CINC in Tampa, it was estimated that the 32 squadrons need to execute the plan would be in place by the end of September. Reynolds, *Heart of the Storm*, 104–5.

37 Gordon and Trainor, *General's War*, 47; Bush and Scowcroft, *World Transformed*, 327–8; Freedman and Karsh, *Gulf Conflict 1990–1991*, 94; Keaney and Cohen, *GWAPS*, Summary Report, 2; Schwarzkopf and Petre, *Hero*, 299–301.

38 Mann, *Thunder and Lightning*, 31.

39 The various troop strength levels, and reference to the briefing to Prince Bandar can be found at: Gordon and Trainor, *General's War*, 40; Freedman and Karsh, *Gulf Conflict 1990–1991*, 88, table 16, 409; Hiro, *Desert Shield to Desert Storm*, 109.

Initial Air Planning: Operation INSTANT THUNDER[40]

From the earliest days of the crisis in August, the potential role of air power was being discussed at the highest levels of the American government. As was the case during Vietnam, opinions varied as to its efficacy. By necessity, the initial military options briefed at Camp David on 4 August relied heavily on air power. When Schwarzkopf and Horner had finished their brief, SECDEF Cheney observed, "The history of air campaigns suggests they are not terribly successful. Why would this one be different?" Schwarzkopf answered by first acknowledging that he was "not an advocate of air power alone, but this is a target-rich environment. There is no cover in the desert."[41]

National Security Advisor Brent Scowcroft, himself a retired Air Force Lieutenant General, also had concerns about the heavy emphasis on air power in this early brief. His main worry was that relying on air power could send the wrong signal to the Saudi's who already had understandable concerns about America's commitment to the region.[42] In truth, at that point the Americans had few other options, and even this one had its risks. As Norman Friedman observes,

> There was no particularly good reason to imagine that a few hundred or even a few thousand tons of bombs, however well placed, would end the problem. There was every reason to expect that any U.S. sponsored air attack would give Saddam Hussein sufficient rationale for invading Saudi Arabia. The United States would then still be faced with a ground war, but one which would begin sooner, allowing little time for preparation.[43]

But it was, for now, the only real choice the CINC had, and he needed to move forward with some semblance of a plan.

In theory and, to be fair, usually true in practice, the planning staff of a regional Combatant Command (COCOM), under the guidance of its commander, is capable of developing an operational plan, briefing it to higher authority, keeping it updated on an established review cycle, and executing it in a time of crisis. But OPLANs are complicated documents and require significant amounts of time and effort to develop effectively. General Schwarzkopf, having just exercised his new draft OPLAN for the first time, had neither the time nor the resources at his

---

40 The code name of the plan was not accidental. Colonel John Warden, its architect, wanted to differentiate his plan from that of ROLLING THUNDER in Vietnam. Reynolds, *Heart of the Storm*, 28–9; Pape, *Bombing to Win*, 220.

41 Bush and Scowcroft, *World Transformed*, 328; Keaney and Cohen, *GWAPS*, Summary Report, 33.

42 Ibid. Earlier in this book, Bush and Scowcroft record the two examples Prince Bandar provides, 325.

43 Friedman, *Desert Victory*, 61.

disposal to accomplish this quickly, and he was very concerned about his inability to offer the president a range of options if asked.[44]

Complicating this situation, his Air Component Commander, Lieutenant General Horner, was in Saudi Arabia overseeing the inflow of American forces, his planners were not coming up with additional options he felt were sufficiently creative, and he seemed to have had little faith in the planning abilities of the other senior Air Force officers on his staff.[45] In what must have been one of the most fortuitous moves he made during the entire operation, Schwarzkopf turned to Washington for help. In response, the Air Force offered up Colonel John Warden, deputy director for war-fighting concepts, and Checkmate.[46]

Checkmate, an Air Force directorate located in the basement of the Pentagon, had originally been formed in the mid-1970s to analyze how the U.S. Air Force would fair against Soviet forces in a major conflict in Europe. It had a unique ability to pull together national intelligence sources, weapons and weapons systems capabilities data, detailed planning information, and experts from a wide variety of fields, to use in conjunction with its classified computers to evaluate air campaigns. The unit was already busy analyzing possible Iraqi targets when the request came in from CENTCOM.[47] At the head of this effort, and the theorist behind the targeting concept for the air campaign in Iraq, was Colonel John Warden.

He was a controversial officer, seen by some as a "better theorist than a commander," in a service culture that values command above almost everything else.[48] In other words, he was a maverick who butted heads with more conventional thinkers, a tendency valued by some and loathed by others.[49] In simplistic terms, during the Cold War U.S. Air Force conceptual thinking, and by extension its force structure, doctrine, and tactics, was primarily based on the categorization of offensive air power into two types, strategic and tactical.

The strategic level was primarily focused on deterrence and, if that failed, taking the nuclear fight to the enemy. At the tactical level the focus was on gaining air superiority near the front and supporting the ground forces. Warden, bucking this generally accepted bifurcated view, thought that a strategic air campaign, using non-nuclear weapons, delivered against properly selected targets, could win a war by convincing the enemy's leadership of the futility of further action.[50] If possible,

44 CENTCOM expected to have the plan completed and approved by the summer of 1991. Keaney and Cohen, *GWAPS*, Summary Report, 29.

45 Gordon and Trainor, *General's War*, 76; Reynolds, *Heart of the Storm*, 8.

46 The details of this extraordinary phone call can be found at Reynolds, *Heart of the Storm*, 23–5.

47 Friedman, *Desert Victory*, 169–70; Gordon and Trainor, *General's War*, 76; Reynolds, *Heart of the Storm*, 15.

48 Gordon and Trainor, *General's War*, 77.

49 Reynolds, *Heart of the Storm*, 30, 77.

50 Gordon and Trainor, *General's War*, 78; Reynolds, *Heart of the Storm*, xi.

then the promises of the early air power advocates, such as Italian General Giulio Douhet, could finally be realized.[51]

He had first outlined his concept for a modern air campaign in a book written while a student at the National War College in Washington, DC. In this volume, *The Air Campaign: Planning for Combat*, Warden identifies five "Air Superiority Cases" which are based on the vulnerability of the respective belligerents' air fields and rear areas, and the relative proximity of the front. The ideal case in his construct is Case II, which is identified as the situation in which your bases and rear areas are safe from attack, the front lines are within reach of your air forces, and the enemy's airfields and rear areas are vulnerable to attack by your aircraft.[52] He wrote:

> A commander must undertake offensive air operations if the battle is to be carried to the enemy, and if political objectives exist beyond hoping the enemy will stop his offensive. It is easiest to envision and discuss offensive operations in their pure state, that is, when every thought can be devoted to the offense without concern for defense.
>
> Case II meets these criteria and is the commander's dream. His bases are nearly immune from enemy attack, but he can attack all parts of his enemy's structure ... Case II provides the opportunity for decisive action—*action so decisive that the war can theoretically be won from the air.*[53]

While Iraq was credited by senior American leaders with possessing a large, battle-tested army, their air force, while numerically large, was seen as "antiquated ... [with] almost no experience [of] using their airpower offensively."[54] It appeared that Iraq just might provide Warden with his ideal Case II, and he planned the campaign accordingly.

Other significant themes emerge from Warden's writings, which also appear to have influenced the planning for the air campaign against Iraq. He was convinced that "command is the true center of gravity," that if an attacking force could separate the commander from his subordinate forces, the "Destruction or isolation of any level of command may have serious—perhaps fatal—impact on the unit or units subordinate to it."[55] Applied to Case II, the apparent scenario in Iraq, he believed that "possibilities are so enormous that the utmost attention should be

---

51 Douhet's theory of victory was, however, slightly different. He believed that if bombers attacked an enemy's populace directly, the resulting outrage would cause a revolt that would end the war quickly. Friedman, *Desert Victory*, 130.

52 John A. Warden III, *The Air Campaign: Planning for Combat* (Washington, DC: National Defense University Press, 1988) Table 1, 21.

53 Ibid., 39: emphasis mine.

54 Bush and Scowcroft, *World Transformed*, 327.

55 Warden, *Air Campaign*, 51, 53.

given to a concerted attack on the enemy's command system."[56] Finally, he warned planners of expected internal opposition, counseling:

> Political leaders may be loath to attack enemy rear areas at times ... It is imperative, however, that the operational commander make clear to the political authorities that they are directing a *militarily* illogical course, and that the cost and duration of the war almost certainly will be far higher and longer than it otherwise might be."[57]

In condensed form, Warden's strategic approach has been called the "five rings" model.

The model can be likened to a dart board, where the bull's-eye (or innermost ring) represents the target of highest value, the true center of gravity, while each of the concentric rings is of lesser importance as one moves outward. It is grounded on the idea that to plan effectively, and thus to get the enemy to bend to your will, you must necessarily view "what is important," through the eyes of your adversary. While the Air Force, for example, generally viewed fielded forces as important to the tactical air campaign, because they were supporting the enemy's ground troops, Warden postulated that what really mattered to the enemy leadership were those things that allowed it to remain in power and command its forces, and that that was what should be targeted.[58]

As a result of this shift in conceptual focus, the center of Warden's model consisted of "leadership" targets, targets whose destruction inhibited "the command, control, communications, and decision-making capability of the enemy." The second ring consisted of "system essentials," the enemy's "military and economic production capability, its factories, electrical grids, power plants, refineries ... all essential to a sustained war effort and modern life." The third contained the nation's "infrastructure," "the means of transportation, movement and distribution ... Destroy these and military and civilian traffic would be paralyzed." The fourth ring was comprised of the enemy's "population." While targeted during World War II, Warden viewed this as a repugnant, non-productive option. Finally, the fifth and outer ring consisted of the "fielded forces."[59]

Warden considered these fielded forces to be in place merely to protect the inner rings and, as such, were the least important targets. It should be immediately apparent that, to the American soldiers who would probably have to face such ground forces on the field of battle, this was a controversial position for him to take, and one which resulted in considerable "push back" to his concept.[60]

56 Ibid., 58.

57 Ibid., 64: emphasis in original.

58 Gordon and Trainor, *General's War*, 78.

59 This summary is drawn from, Gordon and Trainor, *General's War*, 78–9; Reynolds, *Heart of the Storm*, 17; John A. III Warden, *Planning to Win*, Air Power Studies Centre Papers, vol. P66 (Fairbairn, Australia: Air Power Studies Centre, July 1998), 15–16.

60 Warden, *Planning to Win*, fig.7, 17.

In essence, Warden believed that this new conceptualization, along with the improved capabilities brought about by the "military technical revolution," enabled campaigns to be fought in a "parallel" as opposed to "serial" fashion.[61]

In the past, warfare was, necessarily, serial in nature. It required that commanders mass their forces, ground or air, in a given position and take their objectives sequentially, using mass to overcome the lack of accuracy. Now, using precision and his targeting conception, planners could approach the enemy in a systemic fashion. By locating multiple centers of gravity, he also uses the term "target systems," across the "rings," the commander could "bring the enemy under parallel attack, creating a very rapid reduction in the energy level of the opponent and putting it in a position from which it simply cannot react."[62] Thus influenced, Checkmate's planning effort moved forward.

Within about a week this early plan, now conceived as a massive six-day assault of about seven hundred bombing sorties per day, was being briefed to senior leaders. On Friday, 10 August, a team flew to Tampa to brief Schwarzkopf, who was generally impressed. Overriding opposing views from others in the room, he declared, "That's exactly what I want! ... You have my approval—100 percent! ... I will call the Chairman today and have him give you a directive to proceed with detailed planning immediately."[63]

When the issue was raised, he directed that the Air Force's Tactical Air Command (TAC) was not to be given control of the plan, but that Warden and Checkmate were to continue working it.[64] He asked the team to return in a week with more details and to brief CJCS, General Powell, before the following Monday. Powell was, in fact, briefed the following day, Saturday, 11 August, with generally positive results. While declaring that it was a "good plan" and a "very fine piece of work," the general did have two key concerns—concerns that would be raised repeatedly as planning went forward.[65]

First, he asked, what was the plan after the six-day air campaign? Implicit here is the key criticism of "one shot" options. What is to be done next if the first plan doesn't work? The second, reminiscent of the debates in Vietnam, and not surprising given Powell's background, was his grave concern that nothing in the plan was directed at Iraq's forces currently in Kuwait. Even if the plan worked as intended, those forces which had caused so much trouble lived to fight another day. The Chairman told the group, "I won't be happy until I see his tanks destroyed. I want to leave their tanks as smoking kilometer fence posts all the

61 Warden, *Planning to Win*, 8, 15–16. Warden defines the military technological revolution as a "combination of tremendous computational power, increasingly unlimited bandwidth and a number of other technologies—*the key one of which is precision*": emphasis mine.

62 Warden, *Planning to Win*, 16.

63 Reynolds, *Heart of the Storm*, 56.

64 Ibid., 57.

65 Ibid., 71.

way back to Baghdad."[66] While Powell allowed the planning to go forward, albeit with additional guidance, behind the scenes the plan had been facing, and would continue to face, other challenges from inside the Air Force and beyond.[67]

A more detailed version of the plan was briefed to Schwarzkopf on 17 August. The plan was briefed as "a focused, intense, air campaign designed to incapacitate Iraqi leadership and destroy key Iraqi military capability in a short period of time. ... [and] designed to leave basic Iraqi infrastructure intact."[68] The 84 targets then on the target list still did not include any of Iraq's fielded forces. In contrast to both Powell and Schwarzkopf, who both believed this was likely to be a precursor to an eventual land campaign, Warden still believed this plan, alone, would achieve the desired political objectives.[69] That aspect, however, was not focused on during this brief.

Schwarzkopf was happy with the planning effort, and now believed he had a credible response if required, even exclaiming "You've got me so excited with this!"[70] He directed that Warden take the package forward to Saudi Arabia, "I'm sending you to Riyadh, to Horner—to brief him, to hand [the] plan off. My intention is to continue to plan, to refine it to [the] point of execution. ... Don't leave the package, carry and deliver it to Chuck Horner."[71] This simple tasking, delivering the next INSTANT THUNDER brief in Riyadh, was to prove difficult for Warden.

General Horner, forward deployed in the Saudi capital and acting on Schwarzkopf's behalf in the Middle East, had up to this point been almost entirely focused on the never-ending challenges of flowing forces into theater. As CENTAF, and as the CINC's assigned Air Component Commander, Horner quite properly felt that he should have been the one to whom Schwarzkopf had turned to develop the air plan, and he was quite unhappy that people in Washington were involved in "his business."[72] While TAC had forwarded an early copy of the brief on 9 August, Lieutenant Colonel Steve Wilson, one of Lieutenant General Jimmie Adams' trusted subordinates and a man who had worked for Horner before, was sent forward by those opposed to Warden's efforts, to brief the general on the latest version of the plan.

Wilson's brief was delivered on 14 August, and it would be charitable at best to say that the General Horner was not receptive to the ideas contained in it, and the manner in which it was being developed. According to various accounts,

66 Gordon and Trainor, *General's War*, 81–4; Reynolds, *Heart of the Storm*, 71–3; Mann, *Thunder and Lightning*, 63.

67 Reynolds, *Heart of the Storm*, 73–4.

68 Ibid., 86.

69 Ibid., 86–90; Mann, *Thunder and Lightning*, 42, 60.

70 Quoted in, Reynolds, *Heart of the Storm*, 107.

71 Ibid., 110; Schwarzkopf and Petre, *Hero*, 318–20.

72 In fact, CENTAF planners forward had been working on a defensive plan to blunt the expected Iraqi attack into Saudi Arabia when it occurred. Horner deemphasized this aspect in mid-September. Reynolds, *Heart of the Storm*, 116.

he angrily threw the slides across the room not once, but twice.[73] He was also a traditional TAC pilot at heart, and very concerned with the Iraqi armor massed in Kuwait. When the Checkmate team arrived in theater, to brief him on 20 August, the atmosphere had not improved.

The accounts provided by participants who were in the briefing room that day paint a picture of a senior officer "loaded for bear." Horner was imperious and confrontational from the start, even pushing aside a small batch of sundries brought to him by the team at the suggestion of Lieutenant General Adams.[74] Warden, true to his reputation, did not back away from his belief in the soundness of the plan. The colonel and the general talked past each other. Warden felt the general was too focused on the minutiae, while Horner thought the colonel lacked an appreciation of in-theater realities.

Dismissive of the plan's core reliance on striking strategic targets using PGMs dropped from an aircraft, the stealthy F-117A, which was essentially untested in combat, Horner was also unhappy that the plan did not include any provisions for attacking Iraqi ground forces. In agreement with Powell and Schwarzkopf, he did not see this stand-alone plan as a viable option for the overarching realities of the theater. The full plan needed to be longer than six-to-nine days, and it needed to address what he believed to be the very real threat from Iraq's armored forces in Kuwait. It just didn't address enough of the "what ifs?"

Eventually, and perhaps inevitably given the tense environment, Warden crossed the line when he told Horner that he "was being overly pessimistic about those tanks." While the general accepted his quickly offered apology, Warden was effectively marginalized from that point on. A short discussion continued in the room, ending when Horner asked the members of the Checkmate group, except for Warden, to remain in Riyadh.[75]

After the brief, Warden headed back to Washington, henceforth relegated to playing an important but decidedly supporting role for the duration.[76] One of his team, Lieutenant Colonel Dave Deptula, who had actually been working for Secretary of the Air Force Donald B. Rice at the time of the invasion, would remain as a key planner in Riyadh after the other Checkmate officers eventually left, and serve as an important link between the Checkmate office and the air planners forward.[77] With Warden's dismissal, INSTANT THUNDER, with its

73 Gordon and Trainor, *General's War*, 91; Reynolds, *Heart of the Storm*, 91–2.

74 Reynolds, *Heart of the Storm*, 121; Gordon and Trainor, *General's War*, 92.

75 Reynolds, *Heart of the Storm*, 120–30; Gordon and Trainor, *General's War*, 92–3.

76 Reynolds, *Heart of the Storm*, 118–19. By the middle of DESERT SHIELD, Checkmate was "the ad hoc fusion center" for the planning effort. Keaney and Cohen, *GWAPS*, Summary Report, 131.

77 In contrast to Warden, Dave Deptula was a guy "who made few enemies and managed to keep himself in the good graces of superiors and subordinates alike." Reynolds, *Heart of the Storm*, 18.

short duration, PGM-enabled emphasis on strategic targets ceased to exist as an independent plan.

Yet, in an ironic twist, while having been savaged by Horner in public Warden's innovative five-ring approach, and the concepts embodied in the original plan, soon formed the core of the evolving DESERT STORM air campaign. Indeed, Richard Reynolds argues that Warden's dismissal allowed the plan "to move from a rejected idea to bombs on target."[78] Returning Warden to Washington allowed him to use and apply his considerable intellectual skills within the existing Checkmate network to support CENTAF in Riyadh. Deptula, an officer whom both Horner and Glosson worked well with and came to trust, not only provided a direct link back to the Checkmate organization, but had the needed credibility with the two generals to be able to persuade them of the plan's efficacy. And so INSTANT THUNDER would morph into the Desert Storm air campaign.[79]

Finalizing and Executing the Air Campaign in Southwest Asia

With Warden headed back to Washington, Horner now faced the problem of having to find someone to lead the effort to fully develop an effective air plan in Riyadh. He assigned Brigadier General Buster Glosson to lead the highly-compartmented planning team. Glosson was "almost the antithesis of Warden."[80] He, too, was combat tested in Vietnam, but in contrast to Warden he had several successful command tours under his belt. He also got along well with both Horner and Schwarzkopf. Glosson had, until very recently, been serving as the deputy commander of Joint Task Force Middle East, the small naval flotilla in the Persian Gulf, and was more than ready to play a key Air Force role in the current crisis.[81]

The general was first briefed on the INSTANT THUNDER plan on 21 August by members of Warden's Checkmate team. While not fully sold on the plan as a whole, he was interested in its focus on taking the fight into Baghdad from the outset, and not merely focusing solely on Iraq's fielded forces. He did sharethe concerns about what to do next if the six-day plan didn't achieve the desired objectives and of the need to strike the Iraqi ground forces in Kuwait.[82] With that, Glosson's "Black Hole" team began its in-theater planning effort, receiving the first approval of concept from Horner on 26 August.[83]

Over the next several months, the planning group built upon INSTANT THUNDER's conceptual core. In the process, it addressed the significant

---

78 Ibid., 133.

79 A short summary of early air planning is in, Keaney and Cohen, *GWAPS*, Summary Report, 35–8.

80 Gordon and Trainor, *General's War*, 94.

81 Ibid., 94–5; Mann, *Thunder and Lightning*, 65.

82 Ibid., 96.

83 Ibid. Initially the Special Planning Group, the nickname stuck.

challenges presented by: the need to neutralize and defeat the Iraqi Integrated Air Defense System (IADS); the need to coordinate with, resolve doctrinal differences between, and integrate U.S. Navy and allied air assets into the overall air campaign plan; the need integrate and coordinate the use of non-combat-tested Navy Tomahawk Land Attack Missiles (TLAM) and Air Force Conventional Air-Launched Cruise Missiles (CALCM); the requirement to translate broad "target systems/sets" into individual targets with discreet aim points paired with the desired ordnance package, and; the need to develop an air campaign plan that supported all phases of CENTCOM's overall campaign plan.[84]

Although the process was difficult at the best of times, and highly contentious at others, the planners made it work. As the detailed work progressed, the original eight categories of target systems identified in INSTANT THUNDER were expanded to twelve target sets in the DESERT STORM air campaign plan.[85] Glosson and his team worked hard to maintain a dual focus on developing the strategic portion of the air campaign, while simultaneously planning the destruction of Saddam's ground forces. Using a boxing analogy, the Black Hole gang "plann[ed] first to deliver a stunning right cross and then apply a flurry of blows to finish the job."[86]

By mid-September the group had a viable plan that was accepted by President Bush the following month.[87] Constant modifications would continue, of course, until 17 January 1991 and beyond, in light of improvements in intelligence and particularly after the president authorized additional U.S. forces in November.[88] With the authority, contained in UN Security Council Resolution 678, "to use all necessary means … to restore international peace and security" in hand, and following the expiration of the UN-mandated deadline for full Iraqi compliance on 15 January 1991, Operation DESERT SHIELD ended, and DESERT STORM commenced.

When the air war began early on 17 January 1991, the preponderance of the air effort was directed against the strategic portion of the plan.[89] Within a few days, though, emphasis began to be placed on preparing for the coming ground war. There was, however, constant debate between airmen and soldiers over whether or not enough effort was being placed on preparing for the ground war.[90] These

84 Keaney and Cohen, *GWAPS*, Summary Report, 6; Gordon and Trainor, *General's War*, 96–158.

85 Mann, *Thunder and Lightning*, 37–9.

86 Ibid., 80.

87 Gordon and Trainor, *General's War*, 138–9; Keaney and Cohen, *GWAPS*, Summary Report, 39.

88 Keaney and Cohen, *GWAPS*, Summary Report, 6. A tabular presentation of the expanding target list can be found at Gordon and Trainor, *General's War*, 192.

89 H-hour, the scheduled time of the attack, was 3AM (0300 in military time). However, Army and Air Force helicopters attacked and destroyed two SOCs along the Iraqi border 22 minutes before H-hour at 0238. Mann, *Thunder and Lightning*, 117.

90 Hiro, *Desert Shield to Desert Storm*, 336.

tensions even resulted in heated personal exchanges between general officers.[91] Despite this, as history records, the plan worked.[92]

Forty-two days later, "the Iraqi leadership was so disoriented it had to rely on coalition reports for the status and location of its own forces."[93] While overland sorties were flown throughout Iraq, the employment of PGMs had the greatest strategic impact in two areas, in Baghdad and in the Kuwaiti Theater of Operations (KTO). These two sub-theaters will be examined in greater detail.

Air Operations into Baghdad

The opening night strategic attacks in and around Baghdad were swift and effective. As Warden recalled,

> At 1900 hours CNN reported, 'Yes, there are some bombs going off around the city.' We could see that the bombs were hitting where they were supposed to … a few seconds after that I remember turning to the Secretary of the Air Force and saying, 'I'm a bit worried here. It *is* 45 seconds after the war's start and the electricity isn't off. Why isn't the electricity off?'—after all, that was the initial target for the operation. I had no sooner said that the CNN picture went blank.[94]

Using stealthy F-117s, armed with LGBs, and Navy TLAM missiles, the coalition attacked air defense, command, control, and communications (C3), electrical, and leadership targets. This activity, along with a clever follow-on decoy plan, was designed to activate the full Iraqi IADS (called Kari), which it did, thereby leaving it vulnerable to follow-on strikes.[95] The massive blow to Iraq's sophisticated air defense system allowed the coalition to fly with near impunity for the remainder of the war, with few losses.[96]

Throughout the theater, around 1200 strike aircraft flew that first night, hitting more than a hundred of CENTCOM's pre-planned targets, with the loss of only a single coalition aircraft.[97] Again returning to Warden, "within the first 24 hours of the war in Iraq we were able to hit 150 targets of [great] significance ... it simply

91 Gordon and Trainor, *General's War*, 330.

92 By war's end about 15 percent of strikes went towards "strategic" targets, 56 percent against fielded forces. Keaney and Cohen, *GWAPS*, Summary Report, 64–5.

93 Mann, *Thunder and Lightning*, 81.

94 Warden, *Planning to Win*, 5. Bush and Scowcroft, *World Transformed*, 450–51; Gordon and Trainor, *General's War*, 216.

95 Chapter 5 of Gordon and Trainor's *The General's War* is devoted to an examination of this system.

96 Mann, *Thunder and Lightning*, 104; Friedman, *Desert Victory*, 156–8.

97 Keaney and Cohen, *GWAPS*, Summary Report, fig. 5, 13.

put the Iraqis into shock. It was not because the Iraqis were incompetent, but simply because there was no reasonable way to deal with this"[98]

Saddam retaliated asymmetrically, launching a dozen Scud missiles (renamed Al Hussein) into Israel in an effort to bring that nation into the conflict.[99] If Israeli war planes had retaliated, which would necessarily require them flying over Arab nations, there was real concern that such an event would place unbearable tension on the members of the coalition. High level discussions between Washington and Tel Aviv, the shipment of Patriot missile batteries to Israel, and the diversion of some U.S. air effort towards the hunting of mobile Scud launchers served to keep the Israeli's at home, and averted a potential major crisis.[100]

Air strikes into Baghdad continued unabated for the next four weeks. Weather was to prove a negative factor. For example, after two weeks the planners had hoped to have conducted about 300 strikes in Baghdad, yet the actual number was closer to 100.[101] To keep pressure on the Iraqis around the clock, Navy Tomahawk cruise missiles were used for daylight attacks while F-117s were the weapons system of choice at night.[102] The Tomahawk strikes were eventually curtailed after 1 February, in part due to a multi-missile strike during which two missiles were shot down, and in part due to reasonable concerns about the diminishing return on the use of such an expensive missile on strategic systems that were already heavily degraded. While the vast majority of LGB strikes against strategic targets in Baghdad were executed as planned, and with little collateral damage, one incident in particular, the *Al Firdos* bunker raid, highlighted how one incident can significantly alter perceptions, whether the facts or circumstances support a conceptual change or not.

The political fallout from the civilian casualties of the *Al Firdos* bunker raid, to be examined in greater detail shortly, resulted in the CJCS severely restricting F-117 bombing into the capital city after 13 February, although strikes were resumed when the ground war began.[103] By this time, although strategic targets continued to be attacked elsewhere, the focus, particularly at the JCS level was shifting to the coming ground war.[104]

Air Operations into the Kuwaiti Theater of Operations

The initial air operations against Iraq's fielded forces in Kuwait involved older aircraft using decidedly non-precision means, with the kind of less accurate results

98 Warden, *Planning to Win*, 17.

99 Hiro, *Desert Shield to Desert Storm*, 323.

100 Bush and Scowcroft, *World Transformed*, 452–7; Friedman, *Desert Victory*, 19; Hiro, *Desert Shield to Desert Storm*, 323–5.

101 Gordon and Trainor, *General's War*, 312–13.

102 Ibid., 257.

103 Ibid., 408. See also, Keaney and Cohen, *GWAPS*, Summary Report, 24, 69.

104 Ibid., 327. See also, Freedman and Karsh, *Gulf Conflict 1990–1991*, 328.

that would be expected. As Gordon and Trainor note, "The coalition was throwing dumb bombs at the Iraqi troops and doing so from medium to high altitude, tactics that minimized the risk to pilots, but made it hard for the aircrews to see and destroy their targets. Tons of bombs were being dropped, but there were many more misses than hits."[105]

As was the case in Vietnam, there were two obvious non-precision solutions to the problem of attacking an enemy that was hard to detect; to employ mass, again in the form of B-52s, or to use area-coverage weapons. As before, both were tried. Against the fielded forces the Air Force scheduled 24 B-52 sorties per day. But in contrast to Vietnam, and in acknowledgement of Iraq's surface-to-air missile (SAM) threat, only three aircraft were in each bombing cell thus reducing the area covered per mission. More problematic, in addition to battling the high winds in the theater, it was found that a difference in target coordinate systems being used by the B-52s weapons system and that being used by the intelligence planners, later fixed, was resulting in consistent miss distances of hundreds of feet.[106]

The B-52 attacks, while perhaps not terribly accurate at times, had a similar psychological effect on enemy troops in Southwest Asia as that experienced by the North Vietnamese and their proxies in Southeast Asia some two decades earlier. Such attacks posed a real problem for the frontline troops, mostly conscripts, both in terms of their general morale and their very survival. One possible gauge of the level of the bombings effectiveness might be that of Iraqi deserters. While nearly 14 million leaflets urging desertion had been air-dropped on these troops by 6 February, only 87 desertions had been reported. One obvious conclusion that could be drawn is that, at least at this that point, Iraqi morale was quite high.[107] The B-52s began their devastating pounding from above on 7 February, which lead to thousands of desertions before the ground campaign began, and to even more once the operation started. Some of the Iraqi prisoners later reported that the B-52 attacks conducted from 7 February on "were the most devastating."[108]

Tactical aircraft, such as the F-16, A-10, A-6, and A-7 experienced similar difficulties, particularly when trying to accurately deliver "dumb" bombs from high altitude.[109] To overcome such targeting problems, these aircraft compensated using the second approach. Many aircraft used cluster munitions like the MK-20 Rockeye, although there were issues with the standard fuses when used from high altitude, while the Marine Corps resorted to Vietnam era napalm and concussion

105 Ibid., 3. See also, Keaney and Cohen, *GWAPS*, Summary Report, 15.

106 Ibid.

107 Hiro, *Desert Shield to Desert Storm*, 353.

108 Pape, *Bombing to Win*, 247; Keaney and Cohen, *GWAPS*, Summary Report, 15.

109 Higher minimum altitudes were mandated to keep coalition aircraft out of the low-level AAA threat. This made unguided weapons deliveries less accurate. Keaney and Cohen, *GWAPS*, Summary Report, 62.

producing fuel-air explosives.[110] All the while, pressure from ground commanders to focus more on battlefield preparation operations increased.

As described, PGM strikes, both F-117 delivered LGBs and cruise missiles, were having a significant effect on targets in Baghdad, and were also making a largely positive impression on observers, military and non-military alike, around the world. Soon, recent tactical innovations made by pilots in the theater would result in PGMs also having a significant impact on Iraq's fielded forces in the KTO. Tested to a limited degree during the run-up to the war, it was believed that those coalition aircraft equipped with forward-looking infrared (FLIR) systems could effectively search for Iraqi tanks, even those buried in the sand, because the metal of the tanks would absorb more heat during the day, and retain it longer, than the surrounding desert. The theory, in fact, was correct.

First attempted on 5 February, two F-111F aircraft trolled the skies over southern Iraq looking for tanks of the Medina Division of the Republican. When the mission was over, the pair was credited with destroying seven of the eight tanks they engaged with their 500-pound LGBs.[111] "Tank plinking," as the tactic became known, had begun. Glosson's team soon assigned other FLIR-capable aircraft, F-15Es and A-6s to the mission. Assigned to "kill boxes," and aided by "killer scout" F-16 aircraft, efforts to meet the air campaign's assigned goal of destroying 50 percent of Iraq's armored forces had received a strong boost.[112]

The tactic proved psychologically devastating for the Iraqi tank crews. "With nearly all the coalition aircraft operating at medium to high altitude, destruction often seemed to come from out of the blue. Tanks exploded in the middle of the night with no warning" as they were struck by anti-tank missiles or 500-pound LGBs.[113] Armor, a sanctuary for its operators in earlier conflicts such as the Iran-Iraq War, had now virtually become a guaranteed coffin.[114]

In the end, air operations into the KTO have to be seen as successful, although just how successful is open to debate. Unfortunately, the one portion of the *Gulf War Air Power Survey* that might offer valuable insight, a memo from the CIA to the *GWAPS* team titled "Effectiveness of Laser-Guided Bombs Against Iraqi Republican Guard Armor," was deleted from the publicly released version of the survey.[115] Nonetheless, it is widely acknowledged that LGBs decimated Iraqi armor, helped cut the frontline troops off from their supplies within the KTO, killed many Iraqi soldiers, and devastated the morale of those who remained.

---

110 Gordon and Trainor, *General's War*, 319.

111 Ibid., 323. See also, Keaney and Cohen, *GWAPS*, Summary Report, 21.

112 Schwarzkopf and Petre, *Hero*, 431.

113 Mann, *Thunder and Lightning*, 81.

114 Iraqi crews quickly demonstrated that adaptation is still an integral aspect of war, soon abandoning their old sanctuaries and hiding in distant fox holes. Gordon and Trainor, *General's War*, 322.

115 Keaney and Cohen, *GWAPS*, vol. 2, part 2, v.

While the results of different post-war studies vary, there is evidence to suggest that between 84,000 to 153,000 Iraqi troops deserted during the war and that as many as between 30,000 and 36,000 were killed or injured during the air campaign. However, in 1993 *Jane's Defence Weekly* published an article about a report that revised the casualty figures sharply downward concluding that, at the low end of the possible range, perhaps only 1500 Iraqi's had died in battle. This report, John G. Heidenrich's "The Gulf War: How Many Iraqi's Died?" was hotly contested in letters to the editor afterwards.[116] In either case, the number of casualties is quite low, particularly considering the sustained level of effort, when compared to earlier conflicts.

Precision in the Gulf: A Distorted Image?

As alluded to at the beginning of this chapter, during the early part of the air campaign the image of a clean, precise war was televised daily around the world through a carefully orchestrated public relations effort. This comment is not meant to imply duplicity on the coalition's part for it is certainly true that each of us, as individuals and as larger groups, seeks to place the best light on our efforts by accentuating the positive.[117] But there is a significant potential downside, in terms of credibility, if this carefully constructed image is discovered, or deemed, to be false by the audience. For some, it soon seemed that way. As Dilip Hiro observes:

> The early impression engendered by the Pentagon that the allies were carrying out 'surgical strikes' with negligible damage to human life was destroyed when the Pentagon conceded that precision bombs were not being used all the time, that their infra-red sensors were interrupted by rain and fog and smoke, and that not all aircraft were equipped to fire such weapons.[118]

As had to be expected, collateral damage was caused both by the relatively rare errant PGM, and by aircraft that were delivering unguided weapons. And, as in Vietnam, such information, and misinformation, was exploited by various parties to their advantage.

Saddam, for example, attempted to highlight civilian casualties and suffering by letting Western journalists back into Iraq at the end of January with the intent

116 Pape, *Bombing to Win*, 247, 247n.96. The *Jane's* article is quoted in Mann, *Thunder and Lightning*, 123. The original article "Report Puts Iraqi Dead at 1500," *Jane's Defence Weekly* 19 (13 March 1993): 5, notes a Greenpeace study which concluded that as many as 83,000 were killed; Heidenrich's, "The Gulf War: How Many Iraqi's Died?" *Foreign Policy* #90 (Spring 1993), was hotly debated in in later issues.

117 Certainly many viewers were aware that there was a game afoot. Stephen Budiansky caveats a comment about such video with "admittedly selected for public viewing by U.S. commanders in the gulf." Budiansky et al., "Ultimate Test." 30.

118 Hiro, *Desert Shield to Desert Storm*, 336.

of showing them the bombing damage.[119] Under Iraqi supervision, hotels, shops, clinics, and schools were visited by these journalists, struck by coalition bombs the Iraqis said. One report stated that the coalition had struck a bridge in Nasiriyeh in mid-afternoon, when it was full of civilian traffic, killing 47 and injuring another 163.[120] The Iraqi minister for religious affairs, Abdullah Fadel Abbas, and non-Western journalists described attacks on Iraqi critical infrastructure and "mosques and churches ... as well as 80 homes" in a number of cities.[121]

Other journalists filed reports depicting a nation experiencing significant difficulties with its electrical, water, communications, and fuel systems. The American anti-war activist and former U.S. attorney general, Ramsey Clark, wrote of similar difficulties in a report he sent to the UN secretary general in mid-February. Indeed, most of these assertions were, in fact, true and subsequently confirmed by other reporters after the war.[122] But that was just one part of a larger picture of Iraqi society as it existed during the air campaign.

What was more telling, and useful for this analysis, were reports written during the same period that observed that much of the daily life of ordinary Iraqi civilians had remained relatively normal. In newspaper articles published on 1 and 4 February, British journalist Patrick Cockburn reported that,

> Thirteen days of bombing have bred a certain insouciance among ordinary Iraqis. At Abu Ghreib, on the outskirts of Baghdad, a large street market was operating, and there appeared to be as many people in the streets as before the start of the war … In the towns of Ramadi, Falluja and Abu Ghreib, about one in five shops were open … [A] sort of normality has appeared … Food was 'plentiful,' with fruit and vegetables on sale in the shops. Only about a quarter of Baghdad's four million residents had fled.[123]

In another measure of the country's economic state, "prices of candles and eggs had risen fourfold … those of other consumer items such as clothing, shoes, paper products, kitchen utensils, sweets, nuts and spices were virtually unaffected."[124] One reporter, John Simpson of Aviation Week and Space Technology, counted only 29 buildings destroyed in Baghdad, a city of over 4 million.[125] Two important observations are to be made here.

119 Most Western journalists had been expelled from the country at the beginning of the war. CNN's Peter Arnett, and *El Mondo* (Spain) journalist Alfonso Rojo being two exceptions.

120 Hiro, *Desert Shield to Desert Storm*, 353.

121 Ibid., 359.

122 Ibid., 346–58.

123 Ibid., 346.

124 Ibid., 394.

125 Quoted in, Freedman and Karsh, *Gulf Conflict 1990–1991*, 313.

First, while once again not minimizing the suffering and loss experienced by those personally affected by errant weapons, it would certainly seem that such a level of nonchalance or indifference evinced by much of the population, more than two weeks into a devastating air campaign, would indicate that the coalition was, in fact, being quite careful in its target selection and delivery tactics. The fact that reporters were generally describing systemic problems with infrastructure, while Iraqi civilians went about their normal routines, would indicate that the Black Hole planners were actually achieving their stated goal of inconveniencing the Iraqi populace without killing it. Another possible measure of their success is that, despite rhetoric to the contrary, the official Iraqi toll of civilian casualties, killed and injured, on 12 February 1991 stood at 1,397.[126]

The second lesson is primarily for senior decision-makers. Simply put, an image portrayed as part of a strategic communications plan can be remarkably fragile when tested, even when the underlying information on which it is based is factual. For example, on the evening of 13 February 1991, two F-117s dropped GBU-27 "bunker busting" LGBs on the *Al Firdos* bunker in a suburb of Baghdad. One of 25 such facilities around Baghdad, the bunker had originally been built as an air raid shelter but was later reported to have been modified as a command and control facility for Iraqi intelligence personnel, a fact later seemingly confirmed by an Iraqi defector.[127] Others have suggested that the coalition struck the bunker based on a rumor that Saddam slept there.[128] Iraqi Foreign Minister Tariq Aziz denied, after the war, that anyone "close to Saddam" was in the bunker. What exactly he meant by that statement remains unclear.[129]

While the air strike itself was successful, it was soon reported that hundreds of Iraqi civilians, who had been using portions of the bunker as a nighttime air raid shelter, had been killed. This announcement quickly became a public relations nightmare for the US. While the deaths of these civilians, albeit regrettable, would be seen as morally acceptable under the just war principle of double effect, the images of the burned and broken bodies of civilians, including women and children, being removed after the attack ran counter to the public perception of "clean precision" that existed up to this point.[130]

---

126 Ibid., 359. For that period, Iraq finally reported that 20,000 troops had been killed/60,000 injured, 371. After the first day of the air war Iraq reported 89 casualties (23 killed/66 wounded), after four days 232 (41 killed/191 wounded), after 16 days 1078 (428 killed/650 wounded). The most definitive statistics for the entire war, reported by the Iraqi government itself, are 2,278 civilians killed and 5,965 wounded, 325.

127 Bush and Scowcroft, *World Transformed*, 469–70; Gordon and Trainor, *General's War*, 324–7.

128 Hiro, *Desert Shield to Desert Storm*, 362.

129 Pape, *Bombing to Win*, 231n.53.

130 A post-strike review of the tapes from the two aircraft did show the presence of a number of vehicles, including what looked like busses, near the bunker. A review of all available pre-strike imagery confirmed that no such vehicles had been observed there in the past. Gordon and Trainor, *General's War*, 325; Mann, *Thunder and Lightning*, 119.

Based on the information the planners had, the target was a valid target. The impact on future military operations in and around Baghdad, however, was immediate and significant. From that point forward, strikes into Baghdad "would be the exception, not the rule." And, in a manner eerily reminiscent of Vietnam, no targets in the Iraqi capital could be struck without General Powell's personal review and approval.[131] This incident is perhaps a good example of why Ellul was so concerned with the role that "the image" plays in contemporary society.[132]

While Ellul accepted that "images" have existed since the dawn of humankind, he believed a significant shift occurred between 1950 and 1960. While the rise of the television-centric culture is one obvious example, he also used the example of the text book, often now more picture-with-text than text-with-picture to demonstrate the phenomenon is much more pervasive.[133] We have, he was convinced, surrendered ourselves to visual images, largely delivered by television, which have been developed and contextualized by others. We no longer question what lies behind what we see. The image is, or rather has become, reality. Two short excerpts from *The Humiliation of the Word* explore this insight.

In the first, Ellul, while speaking of television's influence on society, observed, "We view a film. We do not perceive an object or reality through our own sight, but rather an image of this reality, seen and encoded by someone else. This image is offered to me as a simple image with no consistency, and my sight causes me to take it for reality itself."[134] The second is similar. Again writing about the television:

> An imaginary mutation takes place that is continually renewed and that erases and takes the shine off reality. A screen of images is placed between me and my world—a circle of images that become so much truer than my own life I cannot rid myself of them. Television is the supremely powerful drug. I end up living my existence before the very thing that eliminates me.[135]

Taken together, his message is clear. In earlier times, humankind spoke to one another and made critical decisions based on such dialogue. Spoken language has as its essential aspects "breadth of meaning, ambiguity, and variation in interpretation." The word is "strictly contradictory to technique in every way," while the image is not.[136]

The challenge for decision-makers is two-fold. First, they must remember their target audience may well be victim to Ellul's concerns, and must build

131 Gordon and Trainor, *General's War*, 326.

132 This event gets almost no mention in Schwarzkopf's memoir. Schwarzkopf and Petre, *Hero*, 435.

133 Ellul, *Humiliation of the Word*, 116.

134 Ibid., 140.

135 Ibid., 120.

136 Ibid., 3, 159.

their communications plan accordingly. If the "image" somehow contains more of the larger narrative that exists behind it, then challenges to that image may have less of an impact than that which occurred in 1991. Secondly, and perhaps equally important, decision-makers must recognize that they, too, are subject to the tendencies described by Ellul and must work to avoid becoming victim to the images presented by others.

An Analysis of the Performance of Precision-Guided Munitions

In part due to their great success during the war, and in part due to the popular mythology that grew up around them, PGMs proved to be the apparent *wunderkind* of Operation DESERT STORM. The 17,000 PGMs used, just 7.5 percent of the total ordnance dropped during the war, caused "well over 75 percent of the serious damage inflicted in Iraqi targets." The American public, due to the nightly clips seen on their televisions, came to see the weapons as "virtually infallible."[137] Trumpeting this success, Warden noted that "In the past, war was defined by misses. This war with precision (the Gulf War) was defined by hits."[138]

The first task then is to determine if these beliefs were accurate, thereby reducing human suffering, and to determine if any other considerations would affect that conclusion. Fortunately, for this effort, the conflict was short, well-documented through a number of media, and extensively debated in writing before, during, and after the war. There are ample resources to examine to try to answer all the frameworks lines of inquiry.

Part I: The Weapons and Their Employment

The first part of this study's analysis seeks to determine just how effective the weapons really were. For the Vietnam case study, although fragmentary, enough declassified data was available to make that process easier. Paradoxically, having just stated how much information is available for academic study in this case study, there is a fundamental problem that even the authors of the *GWAPS* and of the Department of Defense's *Conduct of the Persian Gulf War: Final Report to Congress* acknowledge.

While many weapon deliveries were recorded on aircraft systems for later review, many others were not. This fact, coupled with the highly integrated approach the planners took when dealing with critical targets, means that, as one author observed when speaking of first night strikes, the "raids were so closely interrelated their accomplishments were too interwoven for individual contributions

137 Gillespie, *Precision Guided Munitions*, 203.

138 Warden, *Planning to Win*, 8.

to be readily quantifiable in isolation.[139] Next, much of the specific accuracy data that was obtained was sanitized, i.e. removed or redacted, before the documents were released. Finally, as Iraq was not occupied, as were Germany and Japan, no comprehensive in-country bombing effectiveness study was conducted. The study, then, must rely on information that is in the public domain.

Operational effectiveness In a talk given in Australia several years after the war, John Warden made the following observation:

> The second huge revolution is the military technological revolution which is a combination of tremendous computational power, increasingly unlimited bandwidth and a number of other technologies—*the key one of which is precision*. Those things came together for the first time in the Gulf War where we were able to see what they could actually do as we watched F-117s flying over Baghdad dropping single bombs and *hitting exactly what they were supposed to*. This in my mind is not *a* revolution in military affairs but is *the* first genuine military technological revolution ever because for the first time we now have a *conceptually* different way to wage war.[140]

This statement gets to the heart of this part of the analysis. Simply put, did the weapons really "hit exactly what they were supposed to?" To answer this question, three key categories of PGMs, guided-bombs, air-launched cruise missiles, and the Tomahawk cruise missiles will be evaluated.

Air-delivered guided bombs (LGB/EO) Of the more than 9,000 guided bombs dropped during the war, fully 90 percent were released by the Air Force. The Navy dropped about 7 percent, the majority of which were expended on ammunition storage targets or in the KTO kill boxes.[141]

The Marine Corps accounted for the remaining 3 percent.[142] While the vast majority of air-delivered PGMs were LGBs, some 133 Navy Walleye and a

---

139 Keaney and Cohen, *GWAPS*, vol. 2, part 1, 137n.91; Department of Defense, *Conduct of the Persian Gulf War: Final Report to Congress* (April 1992), Appendix T, T-201.

140 Warden, *Planning to Win*, 8. First two italicized passages added, final two are in the original. This MTR was alluded to earlier (Chapter 2, n.23). For the Soviet/Russian perspectives on the MTR and PGMs, see Mary C. Fitzgerald, "Marshal Ogarkov on Modern Theater Operations," *Naval War College Review* 39.4 (Fall 1986): 6–25; Mary C. Fitzgerald, "The Soviet Military and the New 'Technological Operation' in the Gulf," *Naval War College Review* 44.4 (Fall 1991): 16–43; Mary C. Fitzgerald, "Russia's New Military Doctrine," *Naval War College Review* 46.2 (Spring 1993): 24–44.

141 Schwamb et al., *Desert Storm Reconstruction Report, Vol. II*, 6–13.

142 Keaney and Cohen, *GWAPS*, Summary Report, Table 8, 203. The actual numbers are USAF (8456), USN (623), and USMC (263). Availability of LGB kits was a concern for the Navy's battle force commanders in the Red Sea and Persian Gulf. Frank Schwamb et

few Air Force GBU-15s electro-optically guided weapons were used during the conflict. Additionally, seven new Stand-off Land Attack Missiles (SLAM), a new EO weapon, were fired by the Navy.[143] Recorded images from two of SLAM, as they approached and struck their assigned targets, were used in a Riyadh press briefing.

While little specific data for LGB/EO weapons is available for analysis, comprehensive data was never obtained and those details that were have been removed from public versions of the reports, very positive general comments are contained in the various after action reports. In particular, the *Final Report to Congress*, while acknowledging that weather was an issue at times and that not all weapons guided, stated that "overall, LGBs were responsible for minimizing collateral damage." As for accuracy, the report noted that "Demonstrated LGB accuracy was consistent with results from pre-war testing."[144] Friedman records that PGMs during the war "were credited with a CEP of 1–2 ft, which was probably too high."[145]

Air-launched cruise missiles (CALCM) In the 1970s, the Air Force began development of an air-launched cruise missile, designated the AGM-86, which would increase the capabilities of the B-52 fleet. With the ability to carry several of these nuclear-tipped missiles, each B-52 could strike multiple targets thereby complicating the enemy's defensive problem. While the first variant, the AGM-86A, did not go into production, the follow-on AGM-86B became operational in 1982. Four years later, cloaked in secrecy as a "black program" code-named Senior Surprise, a number of the missiles were modified with a conventional warhead and improved navigation systems. This new AGM-86C model was launched in combat on the first night of Operation DESERT STORM.

To preserve operational security, and to avoid landing in theater with cruise missiles under their wings if they were not launched, seven B-52s flew a 35 hour mission from Barksdale AFB in Louisiana, to the Gulf, and back. Of the 39 CALCMs carried in the flight, 35 were successfully launched against critical targets, while four weapons experienced malfunctions and returned to the United States with the aircraft.

One missile crashed shortly after launch, while 28 were assessed as having hit their targets. Of the other six, three were assessed as "may have impacted the target area."[146] In the *Final Report to Congress*, it was noted that the "Demonstrated accuracy appears consistent with the results obtained from testing."[147] While not

al., *Desert Storm Reconstruction Report, Vol. II; Strike Warfare*, Center for Naval Analyses (Alexandria, VA, 1991), xiii.

143 Friedman, *Desert Victory*, 329–30. Exact numbers do vary slightly.

144 Department of Defense, *Conduct of the Persian Gulf War.* Appendix T, T-183.

145 Friedman, *Desert Victory*, 331. A citation is not provided.

146 Keaney and Cohen, *GWAPS*, vol. 2, part 1, 138.

147 Department of Defense, *Conduct of the Persian Gulf War.* Appendix T, T-179.

publicly available, with an integrated GPS/INS guidance system accuracies should be in the same range as TLAM, which has been publicly reported as being on the order of 15–30 feet, although it may be much better.[148] This first night expenditure exhausted half of the Air Force's inventory of CALCMs, and no other missiles were launched during the campaign.[149]

Tomahawk land attack missiles (TLAM) The Tomahawk cruise missile, a part of the U.S. Navy's shipboard inventory since the 1980s, arrived in theater having never been fired in anger. Because it was untested, and some weren't convinced it was an effective weapons system, planners "were reluctant to give it a major role in the evolving attack plan."[150] More problematic, as one of the potential initial response options available to the theater commander, because Iraq had not previously been considered a hostile state, the missiles did not have any Iraqi targets available to be programmed into their guidance systems, and planning TLAM missions could not be done in theater.[151]

During the war 297 TLAM launches were attempted with 282 of them fully successful.[152] Of the 15 unsuccessful launches, nine failed to leave the launch cell, while six failed to transition to cruise and fell back into the sea. Between two and six were believed to have been shot down.[153] The Navy considered TLAM "fundamentally a success story." As the *Desert Storm Reconstruction Report* records, "A very high proportion of the tasked missions were launched, a very high proportion of the weapons launched appear to have functioned as intended, and a high percentage of TLAM targets were damaged."[154]

Because this formerly classified document was heavily sanitized upon release to the public, there is only limited data to evaluate these statements. But what information is available is very positive. The report notes that of the missiles fired, more than 75 percent were tasked against strategic targets.[155] Of the total fired, 180 were launched within the first 48 hours, with the remaining being launched

148 Carus, *Cruise Missile Proliferation*, 18.

149 Keaney and Cohen, *GWAPS*, Summary Report, 200, 224; Gordon and Trainor, *General's War*, 206, 223. Additional information can be found on the official USAF fact sheet for the missile found at: http://www.af.mil/factsheets/factsheet.asp?fsID=74 (accessed 10 November 2008).

150 James A. Winnefeld and Dana J. Johnson, *Joint Air Operations: Pursuit of Unity in Command and Control, 1942–1991* (Annapolis: Naval Institute Press, 1993), 132; Gordon and Trainor, *General's War*, 114.

151 Gordon and Trainor, *General's War*, 55.

152 A fully successful launch occurs when the missile completes its boost phase and "transitions to cruise," flying its mission profile powered by its onboard engine

153 Friedman, *Desert Victory*, 339; Keaney and Cohen, *GWAPS*, Summary Report, Table 7, 200; Schwamb et al., *Desert Storm Reconstruction Report, Vol. II*, xiv–xv.

154 Schwamb et al., *Desert Storm Reconstruction Report, Vol. II*, xiv, 8-1.

155 A list of the opening night targets can be found in Keaney and Cohen, *GWAPS*, Vol. 2, Part 1, 124.

at an average of seven per day until 1 February. This coincides with other sources that indicate that, after the first night, TLAM was used for daylight missions into Baghdad.

Of the 38 targets assigned, 37 were in fact damaged. While the report redacts exactly how many of the 183 individual aimpoints were damaged, this number can be estimates at around 148 as it is later recorded that "81 percent of the TLAM aimpoints were damaged."[156] Elsewhere the relative accuracy of TLAM is alluded to as it is reported that TLAM hit distillation towers at two refineries at the beginning of the war.[157] As it was with CALCM, the *Final Report to Congress* was complimentary, noting that "TLAM's demonstrated accuracy was consistent with results from pre-combat testing. The observed accuracy of TLAM, for which unambiguous target imagery is available, met or exceeded the accuracy mission planners predicted."[158] While greater detail would have been preferable, it appears that based on the information that is available, as in Vietnam, PGMs indeed were highly accurate and, by all accounts, functioned as designed.

While the raw data available is not as exacting as might be hoped for, the conclusion drawn by the reports' authors seem clear. PGMs appear to have functioned as designed a high percentage of the time, and achieved the low CEPs seen in testing. Therefore, against these two metrics, the trend must be seen as positive.

Manner of employment During an extended explanation of pre-war planning operations, the GWAPS stated, "Though formal restraints by civilian authorities on the military planners were never articulated, in retrospect five can be identified." Three of these restraints are of particular importance to this study. The first two, "casualties among Iraqi noncombatants would be held to a minimum," and a prohibition on striking culturally significant sites, will be analyzed in this section, while the third, "damage to the Iraqi economy and its capacity for postwar recovery would be limited" will be evaluated later when addressing the issue of long-term suffering.[159] While it may be true that these restraints were not well articulated, it is clear that from the very start planners sought to avoid civilian deaths, collateral damage and unnecessary suffering.

The INSTANT THUNDER concept of operations, "called for an absolute minimum of civilian casualties and collateral damage," and, as a result, Warden "excluded direct attacks in the population category."[160] To be truthful, the planners and senior leaders were realistic and knew that some civilians would be harmed. During Warden's 17 Aug briefing to the CINC, while discussing just this topic, "Schwarzkopf interrupted him, a frown beginning to form on his face. 'We must

156 Schwamb et al., *Desert Storm Reconstruction Report, Vol. II,* xv, 8-3.

157 Keaney and Cohen, *GWAPS*, Vol. 2, Part 2, 299.

158 Department of Defense, *Conduct of the Persian Gulf War.* Appendix T, T-202.

159 Keaney and Cohen, *GWAPS*, Vol. 1, Part 1, 90.

160 Reynolds, *Heart of the Storm*, 54.

make folks understand that there will be civilian casualties.'"[161] But the focus was on keeping such unfortunate events to an absolute minimum. This included the decision to only use precision ordnance in areas where collateral damage was highest.

While early air power theorists advocated directly attacking the civilian population to achieve victory, and while attacks on population centers were conducted in World War II and Korea, during this campaign the goal was "to inflict disruption and a feeling of helplessness on the Iraqi public without bringing severe suffering."[162] This emphasis continued beyond just the planning phase. When General Glosson issued his guidance just before the start of the air campaign, he reiterated that "targets and aimpoints will be selected to minimize collateral damage and limit the impact on the civilian population."[163] Air Force Chief of Staff General Ronald R. Fogleman later noted, "while the people of Iraq suffered seriously from degraded services and infrastructure, they remained nearly exempt from direct physical attack. Precise applications of force almost eliminated collateral damage."[164] There is ample evidence that Glosson's directive was adhered to, and Fogleman's observation is correct.

On the first night of the war, 18 F-117s did not drop their bombs due to bad weather, choosing to return to base rather than drop them unguided.[165] In fact, F-117 pilots remained well disciplined when making their final drop/no drop decisions over their assigned targets, which were often in heavily populated downtown Baghdad. As with any other LGB delivery platform, their FLIR/laser systems were very vulnerable to bad weather as a clear line of site between the aircraft and target must be maintained for accurate delivery. As a result, "only 60 percent destroyed their targets, although this was a high proportion of those actually launching their weapons."[166]

That other pilots also maintained the discipline needed to keep collateral damage low was evident on the second night of the war. A large package of F-16s, for example, was sent to attack the Tuwaitha nuclear research facility on the edge of Baghdad. Under attack by ground defenses, and thwarted by poor weather and Iraqi smoke pots designed to obscure the target, many of the aircraft chose not to drop their ordnance.[167] Such discipline helps to explain why in "the first 36 hours of the

161 Ibid., 103.

162 Keaney and Cohen, *GWAPS*, Summary Report, 43.

163 Freedman and Karsh, *Gulf Conflict 1990–1991*, 312. Guidance issued on 12 Jan 1991.

164 Mann, *Thunder and Lightning*, ix.

165 Gordon and Trainor, *General's War*, 220.

166 Freedman and Karsh, *Gulf Conflict 1990–1991*, 313.

167 Gordon and Trainor, *General's War*, 256.

air war, numerous targets were attacked in and near major Iraqi cities, and—even according to Iraqi claims—only 23 civilians were killed."[168]

The Iraqis, as the Vietnamese had earlier, recognized the self-imposed restraints of the Americans and tried to take advantage of them. Similar to the AAA sites that were placed on dikes in North Vietnam, Saddam reacted to the US concern with civilian casualties by placing military equipment on and around sensitive civilian sites such as schools, hospitals, and private homes to avoid their destruction.[169] As before, aircrews maintained their discipline.

Finally, while certainly in part for political and national image reasons, the president and his advisors became concerned that the coalition not be seen as killing retreating troops needlessly on the "Highway of Death" between Kuwait City and Basra. Although possessing the technological means to destroy more of Saddam's forces, with little risk to coalition forces, the decision was made to end the war.[170] What is perhaps more telling, as the fog of war lifted, and it was clear that not as much Iraqi armor had been destroyed as had been believed, the choice was made not to reengage the battle. Evidence suggests that such mitigation efforts were largely successful.

From a decidedly non-military source, William Arkin, a political commentator and activist working at Greenpeace at the time, provides some interesting insights. In a 31 October 1991 brief to the *GWAPS* team, he concluded that 2,278 Iraqi civilians had been killed and 5,976 were injured by coalition bombing during the air campaign.[171] Most importantly, having himself personally inspected many of the strategic targets after the war, he declared that, "the air war was 'clean on a strategic level,' and that he could find no evidence of indiscriminate attacks on cities or civilians … or extensive collateral damage of civilian structures near targets."[172]

Middle East Watch, a division of the Human Rights Watch organization, conducted an extensive examination of the effects of the air campaign on civilians. The study, while striking a tone diametrically in opposition to that of the military and civilian leaders at the time, nonetheless arrives at numbers similar to Arkin's. The study concludes that, at the upper limit of their estimated range, 2,500 to 3,000 Iraqi civilians perished in the air campaign throughout Iraq.[173] Again, while the following comment is not meant to downplay the human suffering that lies behind

---

168 Blackwelder, *Long Road*, 25. Original source Stephen Budiansky et al., "Air Power's Ultimate Test," U.S. News and World Report, vol. 110, no. 3 (28 January 1991): 31–2.

169 Freedman and Karsh, *Gulf Conflict 1990–1991*, 325.

170 Bush and Scowcroft, *World Transformed*, 485–6.

171 Quoted in Keaney and Cohen, *GWAPS*, Summary Report, 75n44.

172 William Arkin, "Tactical Bombing of Iraqi Forces Outstripped Value of Strategic Hits," *Aviation Week & Space Technology*, 27 January 1992, 62.

173 *Needless Deaths in the Gulf War: Civilian Casualties During the Air Campaign and Violations of the Laws of War*, (New York: Human Rights Watch, 1991), 19.

these numbers, these are remarkably low figures when compared to previous wars, although not dissimilar to those seen in North Vietnam during LINEBACKER II, when PGMs were heavily utilized, an interesting fact of its own.

Against the framework's two metrics, the trend lines can be viewed as positive. There is clearly sufficient evidence of deliberate planning aimed at reducing casualties and collateral damage. This emphasis was in place from the beginning of the INSTANT THUNDER plan at Checkmate, and remained present throughout. Similarly, while perhaps not as well documented, there is clear evidence that certain targets, cultural sites and mosques among them, were placed off limits. That even anti-war writers would acknowledge the success of these efforts is significant.

Part II: The Question of Unintended Consequences

The question of unintended consequences can be explored in depth with this case study. Elements of all three lines of inquiry were evident in Iraq. Most interestingly, perhaps, this case provided an unanticipated insight into the first line of inquiry. As will be seen, while the metric initially developed was fundamentally inward facing, the war in Iraq demonstrated that unresolved tensions can also contain an outward-facing component as well.

Rapid capitulation with unresolved tensions When this portion of the framework was originally developed, the situation envisioned was that of one nation engaging another on its own territory. The expected display of "unresolved tensions," hence the metric selected, was the emergence of something like an organized résistance movement rallied against the victorious belligerent. This would be, fundamentally, an inward phenomenon. That an insurgency occurred after the war ended, both in the north and the south portions of the country, is well known. But the fact that the insurgency was directed at Saddam is significant. While it seems clear that many in the coalition would have been happy had Saddam been toppled from power, it was never an explicit, agreed upon coalition objective. Thus, while unresolved societal and political problems certainly remained, the additional human suffering involved cannot be tied back to the use of PGMs. No trend line analysis is possible as such suffering would still most likely have occurred had the war ended the same way, with only "dumb" ordnance having been used.

In this case, however, another manifestation of unresolved tensions emerged, that of Iraq's relationship with its regional neighbors. If the coalition only got Iraq out of Kuwait, but left the majority of its armed forces intact, it effectively would have only pushed another regional crisis into the future. And, most importantly, the initiative would have remained with Saddam. Neither the United States, nor its coalition partners, could handle the open-ended regional commitment such a situation would require.

It is clear that the Bush Administration recognized this and worked hard to resist calls from critics who would enable such a state of affairs. The Soviets, for example, called for a cessation of combat operations after the first night of the war. Bush countered Mikhail Gorbachev's pleas saying that stopping so soon, right after Iraq had launched Scuds into Israel, and with most of its military untouched, would allow Saddam Hussein to claim victory in the region.[174] Cheney and Scowcroft held similar views, that "a ground campaign would be necessary no matter what air power accomplished, because it was essential that we destroy Iraq's offensive capability."[175] With the U.S. in the lead, the coalition pressed forward with the ground war and, while perhaps not destroying as much as had been believed, dealt Iraq's military a significant blow which reduced the threat in the region. In addition to satisfying one of the four key objectives, it can be seen as a positive trend line.

Adverse asymmetrical responses Unable to counter the coalition assault on his capital city and fielded forces, Saddam in fact sought to influence the outcome of the conflict through a number of asymmetric means. The first, previously discussed, involved the launching of Scud missiles into Irael. While certainly a political move designed to bring Israel into the conflict, thereby possibly breaking the coalition, such attacks also had the potential for causing a great degree of human suffering, particularly if the warheads contained chemical or biological agents.

In another example of an adverse asymmetrical response, following the *Al Firdos* bunker incident Saddam Hussein publicly announced that "in the future, *all* Iraq's military bunkers would be used as well by civilians."[176] Unable to counter the coalition's air assault in Baghdad, highly effective due to PGMs, he appeared willing to stop such strikes by using what were essentially civilian hostages, or by creating a series of ongoing public relations nightmares using the bodies of his own people. As noted earlier, the gamut worked.

A third example of an adverse asymmetric response involved oil pipelines near the coast. Early in the war, on or about 25 January 1991, Saddam ordered the release of millions of gallons of crude oil into the Arabian Gulf. In addition to the basic ecological damage that was being done, the slick held the very real possibility of drifting into the intakes for the coastal desalinization plants, including the one at *al Jubayl*, the world's largest. Even shutting down the plants, to avoid contamination, could have caused significant suffering due to the loss of fresh water if required for an extended period of time. Fortunately, favorable winds and currents kept the oil away from the plants and sensitive marshlands.[177]

---

174 Bush and Scowcroft, *World Transformed*, 454.

175 Ibid., 463.

176 Ibid., 454.

177 William M. Arkin, "On Impact: Modern Warfare and the Environment: A Case Study of the Gulf War"

A Greenpeace study prepared for a "Fifth Geneva" Convention on the Protection of the Environment in Time of Armed Conflict (3 June 1991, London, UK), 62–6.

Of note, as the oil slick spread it was determined, with the aid of oilfield experts, that the destruction of two critical pipeline manifolds would stem the flow of the oil. On 26 January 1991, F-111F aircraft used GBU-15 (2000-pound) EO weapons to destroy the components.[178] Thus, the very type of weapon that might have caused the asymmetric response, actually helped avert a major ecological disaster.

These three, coupled with the torching of the Kuwaiti oil wells, could have caused significant human suffering if the course of events had been different. While, in this case, most suffering was avoided, it must be acknowledged that the trend line could easily have been negative.

Long term suffering From the beginning, the third of the implicit restraints listed above, that "damage to the Iraqi economy and its capacity for postwar recovery would be limited," was explicitly addressed by the military planners in Checkmate. In Warden's opinion, "the [INSTANT THUNDER] plan would destroy Saddam Hussein's power base and leave his offensive military capability degraded and difficult to rebuild. It would also severely disrupt Iraq's economy. Unlike the postwar military, however, Iraq's economy could be quickly restored."[179]

These concerns were openly discussed at the highest levels. During the 10 August briefing to Schwarzkopf, Warden "cautioned against destroying Iraq's oil production capability for fear the nation would not be able to pay any war debts or rebuild itself after hostilities."[180] At the second briefing in Tampa, one briefer, Colonel Blackburn, talked about "shutting down the cracking towers at the *Al Basrah* and *Az Yubayr* petroleum refineries." In doing so, the intelligence officer reminded the CINC, "the air campaign would reduce or stop the flow of oil internally in Iraq yet allow quick recovery of the industry at war's end, thereby causing little or no harm to Iraq's export capability."[181] Mann records a similar view of the planning process:

> Because of a U.S. commitment 'to the security and stability of the Persian Gulf … planners did not wish to cripple Iraq's ability to maintain a balance of power in the region after the war. Thus, instead of destroying the country's means of producing and exporting crude oil, coalition airpower would target only internal pumping and storage facilities. This action would affect domestic distribution and usage but would allow a rehabilitated Iraq to reestablish exports after it complied with coalition (i.e., UN) demands.[182]

178 Gillespie, *Precision Guided Munitions*, 204; Gordon and Trainor, *General's War*, 266.

179 Reynolds, *Heart of the Storm*, 105.

180 Ibid., 54.

181 Ibid., 104.

182 Mann, *Thunder and Lightning*, 41.

With regards to the oil industry, this concerted effort at avoiding/reducing long-term human suffering appears to have been successful. Using numbers provided by Iraqi officials, the Gulf War Air Power Survey concluded that "there is no evidence that the air campaign inflicted lasting infrastructure damage to Iraq's oil industry."[183] A similar logic was also used when devising the plan to turn the lights off in Iraq.

That portion of the overall campaign plan was designed so that, in theory, the system could be rebuilt in months rather than years. This process included a careful analysis of the system. For example "the planners reasoned that attacking the transfer stations (transformer yards) would have the same effect as striking the generator hall (i.e., disruption of the electrical power grid) but would be much easier to repair after Iraq complied with UN demands."[184]

Unfortunately, it must be admitted that not all of the pilots adhered to the concept, in part due to the fact that the verbal policy on electrical installations had not been passed to all in the chain-of-command.[185] When this was discovered, in early February, a memorandum was prepared by Deptula for Glosson to issue. It stated,

> Electrical targets will be targeted to minimize recuperative time. At electrical production/transformer stations the objective will be the transformer/switching yards and the central buildings in those yards. Boilers and generators will not be aimpoints. The objective at POL targets is to destroy the refined product storage. Distillation and other refining areas will only be an aimpoint for military related fuel production refineries.[186]

By this time, however, much of the damage the planners deliberately sought to avoid had already been done to the electrical generation capability of Iraq. In the fog and friction of war, a humanely conceived policy was inconsistently followed. As William Arkin correctly observes, "You can't separate neat and clean bombing from postwar deaths, people just died in a different way because of the efficiency of the attacks."[187] As a result, a degree of responsibility for increased human suffering must be assigned in this case, with the trend line seen as negative.

A post-war survey conducted by the Harvard University School of Public Health asserted that damage to the electrical system "hampered efforts to purify water and contributed to an alarming escalation in infant mortality rates."[188] Because Iraq relied heavily on electrical power for water purification and distribution,

183 Keaney and Cohen, *GWAPS*, Summary Report, 76.

184 Mann, *Thunder and Lightning*, 41.

185 Keaney and Cohen, *GWAPS*, Summary Report, 72.

186 Quoted in Gordon and Trainor, *General's War*, 316. The memo was backdated to early January.

187 Arkin, "Tactical Bombing of Iraqi Forces," 62.

188 Gordon and Trainor, *General's War*, 316–17.

and sewage removal and treatment, the loss of power had significant negative effects.[189] Finding "sharply increased levels of gastroenteritis, cholera, typhoid, and malnutrition in Iraqi children due to the delayed effects of the Gulf War," the study estimated that there were 70,000–90,000 additional deaths, due to increased mortality rates, between April-December 1991.[190] William Arkin estimated that 110,000 additional Iraqis eventually died due to the "indirect detrimental health effects" of the war.[191]

These medical conditions outline just a portion of the problems caused by the loss of, or reduction in, reliable electrical power. Agricultural production, due to loss of irrigation systems was also affected. Refrigeration capability was lost, impacting the transportation and storage of perishable foodstuffs. In addition to an increase in waterborne illness, the Iraqi healthcare systems, particularly its hospitals were negatively impacted.[192] Had the bombing plan been executed as conceived, much of this suffering might have been averted.

It is also evident that there were many other sources for the prolonged human suffering experienced by many Iraqi citizens after the war. Seeing what was happening, on 7 March Iranian President Rafsanjani told Saddam, "You know well that you are undesirable in your country as well as in the region. So, don't further stain your bloodied hands by killing more innocent Iraqis. Yield to the people's will and step down."[193] This fell on deaf ears and the response was brutal. The southern marshes, once home to some 500,000 people, were destroyed. Uprisings by the Kurds in the north, and Shia in the south were brutally repressed by the regime, with the loss of thousands of lives. The coalition simply watched, as older regional concerns about the Shia in the south and the Kurds in the north reemerged.

While it has been charged that the Bush Administration bears some moral responsibility for having called on the Iraqi people to rise up and overthrow Saddam, such a position is not related to the use of precision-guided munitions. Rather, these were political issues of the highest order. Such an accusation must also be balanced against the question of what would have happened if the U.S. had altered its political objectives. Putting aside the very real issues of coalition cohesion and world opinion, particularly on the "Arab street," SECDEF Cheney captured the essence of the issue when he remarked on BBC radio,

> If we'd gone to Baghdad and got rid of Saddam Hussein—assuming we could have found him—we'd have had to put a lot of forces in and run him to ground some place. He would not have been easy to capture. Then you've got to put a new government in his place and then you're faced with the question of what kind of government are you going to establish in Iraq? Is it going to be a Kurdish

189 *Needless Deaths in the Gulf War*, 172.

190 Quoted in Keaney and Cohen, *GWAPS*, Summary Report, 75n.45.

191 Ibid.

192 *Needless Deaths in the Gulf War*, 177–86.

193 Hiro, *Desert Shield to Desert Storm*, 400.

> government or a Shia government or a Sunni government? How many forces are you going to have to leave there to keep it propped up, how many casualties are you going to take through the course of this operation?[194]

Additionally, much of the degradation in post-war Iraq was needlessly caused by Saddam's refusal to fully comply with the UN mandates necessary for an easing of the trade embargo. Similarly, his decisions as to how to allocate the scare resources he had at his disposal contributed to his people's suffering.[195] Choosing to rebuild symbols of power, such as destroyed palaces, rather than damaged infrastructure is but one well known example. All of these are tragedies to be sure, but remain separate from this analysis.

Part III: The Question of Teleology

Whereas the Vietnam case study was difficult to evaluate at the teleological level with regards to PGMs, because the technological capability arrived after the war started, this case study illuminates this portion of framework particularly well. Having the ability to use PGMs, perhaps exclusively, was discussed throughout all phases of this operation. It is, to be sure, a seductive option if it could work, but examination along the two lines of inquiry demonstrate why such an option was not selected.

Reluctance to deploy conventional land forces When looking at this particular line of inquiry, it is perhaps insightful to look at what is meant by the word "deploy." It could mean, on the one hand, the sending of troops forward from their peacetime bases. The Navy, for example, calls extended cruises "deployments" even when no combat operations are anticipated. In another sense, it might be countered that "deploying the troops" means sending them onto the field of battle from their forward staging bases. In either case, there seems little to support a charge that President Bush was reluctant to deploy ground forces.

In fact, what is interesting in this case study is that while some military officers, Colonel John Warden among them, believed that an air campaign alone could win the war, the senior leadership, both civilian and military was not drawn to this, what might pejoratively be termed, "cheap option."[196] From as early as the end of August 1990, it was clear that the Bush Administration was taking the long view. The political, diplomatic, and military moves during those first four weeks were "intended to keep Iraq isolated and deter it from future aggression until it either bowed under political and economic pressure or we had

194 Freedman and Karsh, *Gulf Conflict 1990–1991*, 413.

195 See Fran Hazelton ed., *Iraq Since the Gulf War: Prospects for Democracy* (London: Zed Books Ltd., 1994).

196 In contrast, Congressman Les Aspin (D-WI) felt air power alone could win the war in about three weeks. Friedman, *Desert Victory*, 133.

the military means and international political support to use force," force which likely involved a ground force.[197]

Not that there weren't some concerns at the highest levels. Following the 10 October briefing of the overall air and land campaign plan to the Chairman, one day before the White House was to be briefed, Powell was clearly concerned that "the White House might act on the plan—or at least the air campaign portion" believing the "air campaign might be too attractive."[198] Glosson was directed not to "oversell" the plan, or make "the air campaign look too easy."[199]

The question of whether the air campaign alone could force Saddam out of Kuwait was asked at the 11 October briefing to the National Security Council (NSC) core group. Powell's cautious response was that it would probably not be enough.[200] President Bush seems to have grasped this reality and, while certainly not looking forward to ordering troops into combat, showed no tendency to avoid using ground forces merely because of the air component's precision capability. In fact, at the 30 October NSC meeting, based on military advice, the decision was made to further increase the level of U.S. troops beyond the 250,000 currently in, or headed to, the theater.

This was a critical decision, for which the president was heavily criticized in some circles. By authorizing this increase, the US had taken the possibility of troop rotation off the table.[201] Even after the air war had begun, and was apparently achieving great success, the senior leadership resisted calls to forego a ground war. In fact, the greater concern expressed by President Bush was that the air campaign would work, or a political solution would be found, before they had a chance to deal his ground forces a final blow.[202]

Finally, President Bush's strong resolve in the face of significant pressure by Gorbachev in the final hours before the ground demonstrates that, while not eager to place coalition troops in harm's way, he was not willing to continue negotiating with what he considered an insincere Iraq while people still suffered in Kuwait.[203]

Greater willingness to resort to lethal force It might be argued that having an impressive plan for a strategic air campaign, and the technological means to carry it out, might predispose the American senior leadership to use force more readily. The president was certainly aware of his military's capabilities. In remarks made at the Aspen Institute, as the crisis in the Gulf unfolded, President Bush noted,

197 Bush and Scowcroft, *World Transformed*, 355.

198 Gordon and Trainor, *General's War*, 134.

199 Ibid.

200 Bush and Scowcroft, *World Transformed*, 380; Gordon and Trainor, *General's War*, 135–41.

201 Ibid., 393–7, 400–402.

202 Ibid., 466.

203 Ibid., 476–7.

> Let me begin with the component with great long-range consequences: research. Time and again, we have seen technology revolutionize the battlefield. The U.S. has always relied upon its technological edge to offset the need to match potential adversaries' strength in numbers—cruise missiles, Stealth fighters and bombers, today's "smart" weapons with the state-of-the-art guidance systems, and tomorrow's "brilliant" ones. The men and women in our Armed Forces deserve the best technology America has to offer.[204]

But it is also clear, particularly from the personal reflections contained throughout *A World Transformed*, that President Bush was in no rush to go to war and have American blood shed on the battlefield, even with his high tech advantage.

As earlier noted, he recorded in his personal diary, "The thing that weighs on me is sending kids into battle and the lives being lost." [205] In a letter to his five children on the last day of 1990, he wrote:

> Every human life is precious. When the question is asked "How many lives are you willing to sacrifice"—it tears at my heart. The answer, of course, is none, none at all ... we have moved a tremendous force so as to reduce the risk to every American soldier if force is to be used; but the question of loss of life still lingers and plagues the heart.[206]

That this belief that life was precious extended beyond merely the American soldier is also reflected.

In a diary entry written just before the UN-mandated time limit expired he wrote, "The reports from Baghdad are defiant. People marching in the streets—turning out. Yet their faces smile and they chant. And I think, Oh God, save their lives."[207] Even on the eve of the ground war Bush's thoughts returned to the troops. In his final diary entry before the ground war commenced he wrote, "I've got a deep concern about the families, the loss of lives, and am wondering how it will all end."[208]

But this case study also serves as a reminder that America's founding fathers placed decision-making power in the hands of several key branches of government. As the military was planning, and President Bush and his team were working diplomatic and political channels, Congress was seeking to ensure that checks and balances were in place. At a 30 October meeting with Congressional leaders, for example, Bush was handed a letter signed by 81 Democrats which, in part, stated that they were "emphatically opposed to any offensive military action."

---

204 George H.W. Bush, President, *Aspen Institute Symposium*, Speech (Aspen, CO: Aspen Institute, 2 August 1990).

205 Ibid., 381.

206 Ibid., 435.

207 Ibid., 449.

208 Ibid., 479.

Furthermore, if the president later determined force was required, "he must seek a declaration of war from Congress."[209] While the trend line is positive, as there is no evidence of a greater willingness to use force, it is also worth noting that, for the American system, such willingness may well be checked by cooler heads.

Returning to the sub-theme of the epigraph at the beginning of this chapter is perhaps a good point to begin the summary. As Robert Pape observes, "People's views on the future of coercive air power are likely to be determined by their interpretation of the Persian Gulf War. The many televised films of modern precision-guided weapons making deadly accurate strikes on all types of targets have fed a perception that a technological revolution has made it possible to win wars with air power alone."[210]

As previously noted, this theater was perfect for airpower's employment.[211] As Friedman observes, "The flat and largely empty terrain is practically made for air attack; an army literally cannot hide. It can reduce its vulnerability, but the war showed that precision-guided weapons largely negated classic measures like digging in."[212] Even so, the first week of the air war, what Hiro terms "the most intensive aerial bombing in history" did not, as Warden had believed, force Iraqi capitulation and end the war.

Iraqi resiliency and innovation frustrated "US Air Force expectations that six days of intensive bombing would destroy the [Iraqi] air force, ballistic missiles and command system, [thereby] bringing Iraq to its knees, or at least making an early ground battle for Kuwait an easy option with minimal casualties."[213] In actuality, the extended air war also did not result in Iraq's capitulation. The INSTANT THUNDER plan rested on the fundamental assumption that "Saddam's regime rested on its physical ability to provide secure leadership areas and telecommunications to support forces that were vulnerable to air attack. *In fact, it rested on a political structure that air attack could not alter*."[214]

During the 1991 Gulf War, the American forces held a significant technological advantage over their adversary, in a theater practically tailor made to exploit it—Warden's ideal "Case II." Yet, the trend lines, across all three parts of the evaluative framework remained largely positive. The planners took human suffering concerns seriously and the weapons worked as designed. PGMs did not add to the long-term suffering experienced post-war, while senior leadership, carefully considering all aspects of the situation, did not reach for the easy solution some offered. In Iraq, "the Coalition forces enjoyed the inestimable advantage of operating against a

209 Ibid., 389; Freedman and Karsh, *Gulf Conflict 1990–1991*, 290–95; Bush and Scowcroft, *World Transformed*, 446.

210 Pape, *Bombing to Win*, 211.

211 Skolnikoff, *Elusive Transformation*, 73.

212 Friedman, *Desert Victory*, 237.

213 Quoted in, Hiro, *Desert Shield to Desert Storm*, 335. Original is from an article in the 24 January 1991 issue of the British newspaper *Guardian*.

214 Pape, *Bombing to Win*, 226: emphasis mine.

hopelessly exposed opponent caught in completely flat, open terrain. Future wars will not necessarily follow that pattern."[215]

What was apparent was that the focused targeting of strategic targets, using PGMs, had achieved a new level of attention within the senior leadership, which could influence future operations. Gordon and Trainor are also concerned that DESERT STORM created new expectations that casualties, on both sides, would be low in all future conflicts. Having been willing to sustain tremendous casualties when the cost seemed necessary, World War II being one example, planners have become overly focused on keeping casualties low, they believe. They also believe that the American public has become too detached from the realities of combat, and that our future enemies will seek to take advantage of this fact.[216] The next case study validates some of these concerns.

215 Van Creveld, *Technology and War*, 269.

216 Gordon and Trainor, *General's War*, 469–70.

Chapter 6
The Balkans: Perfect Precision Attained?

> Dumb bombs are dead!
>
> Lieutenant General Michael Ryan, USAF

> It does now look as if air power has prevailed ... the light began to dawn because we (the military analysts) had gravely underestimated the significance of bombing with modern precision weapons.
>
> Sir John Keegan, Daily *Telegraph*[1]

As indicated by the two epigraphs above, the use of PGMs in the two Balkan campaigns was a watershed event for this particular technology. The long history of the peoples of the Balkan region, a complex admixture of ethnic and religious tensions, periodic fighting, and persistent memory, is quite difficult for an outsider to readily comprehend. With the collapse of federalist Yugoslavia in 1991, these latent tensions were once again released bringing more than a decade of civil war and ethnic fighting to the Balkan region. While, as R. Craig Nation points out, this decade also witnessed a number of other, equally devastating regional conflicts around the world, Rwanda, Sierra Leone, Somalia, and Haiti prominent among them, the Balkans captured the West's attention. As a part of Europe, although seen by some as a "grim backwater," the region was readily accessible to the media.[2] As a result, this case may serve, perhaps, as a counter-argument to Ellul's observation that humankind's fixation with the image is a negative development.

While in the previous case study the image of perfect precision actually masked a more complex truth, here the images of human suffering broadcast worldwide may have served as a catalyst for a desire for greater understanding, and action. As Nation writes, atrocities in the Balkans were "prominently displayed and extensively discussed on television news reports. The resulting impact on elite and public opinion made the Balkan conflict politically compelling—it was a war that could not be ignored."[3] Diplomatic attempts to resolve the internal conflicts, and bring the fighting to an end, eventually involved international organizations such as the UN, the EC (later the EU), NATO, along with the efforts of the new Russian

---

1 The former epigraph is from a post-Bosnia interview, the latter from a post-Kosovo *Daily Telegraph* (UK) editorial. Keegan had harshly criticized the action earlier. He soon backed off from this proclamation however.

2 R. Craig Nation, *War in the Balkans, 1991–2002*, monograph (Carlisle, PA: Strategic Studies Institute, 2003), vii.

3 Ibid.

Federation and the U.S. Ultimately, however, these efforts were unsuccessful and military action was required.

On two occasions, intensive air operations, Operation DELIBERATE FORCE (1995) in Bosnia and Operation ALLIED FORCE (1999) in Kosovo, were instrumental in bringing the belligerents to a diplomatic settlement. These two operations were impressive applications of combat force, largely prosecuted using precision-guided munitions. In fact, compared with less than 1 percent in Vietnam, and around 8 percent in Operation DESERT STORM, in Bosnia over 90 percent of munitions used were precision-guided, although for reasons that will be explained the percentage dropped again in Kosovo.[4] As a result, the operations were hailed by air power advocates as final vindication of the early theorist's belief in victory through air power alone.

Sir John Keegan, the noted British military historian and author, trumpeted as much from the pages of the *Daily Telegraph*.[5] But is this image accurate? The study will examine these two operations, which were almost entirely devoid of associated allied ground combat troops, to determine what effect PGMs actually had on human suffering in each, and how the availability of PGMs influenced key decision-makers. As before, a three-part framework will be utilized.[6]

The Balkans 1980–1995: An Abbreviated Overview

The history of the Balkan's region is difficult for the outsider to comprehend, and no attempt will be made here to rectify that situation. Rather, the study will take as a given that the rich and complex history of the region has been greatly influenced by the rise and fall of empires, of autonomous feudal principalities, and by the associated ethnic and religious tensions that often times occur when differing peoples are forced to live in close proximity. Ultimately, fully understanding the specific historical events behind this statement is unnecessary for this effort. While critics may claim that by doing so this study will miss some key historical aspect of the case study, the focus of this work is on a technology, specifically PGMs, which was only coming to prominence in the Gulf War at about the same time as the Yugoslav federation began its final dissolution.[7]

4 Estimates of between 94 and 98 percent can be found in the literature.

5 Keegan was not alone. In a 7 July 1999 article Nick Cook called Kosovo "arguably the first time in history that a conflict has been won using air power alone." Nick Cook, "War of Extremes," *Jane's Defence Weekly* 32.1 (7 July 1999): 21.

6 There were UN peacekeeping forces (UNPROFOR) in the region at the time of DELIBERATE FORCE, but these troops did not have sufficient authority, combat power, or support to intervene effectively.

7 For an excellent primer on the history of the Balkans, see Nation, *War in the Balkans*, 1–32.

Marshall Josip Broz Tito ruled Yugoslavia for nearly four decades. First coming to prominence as one of the partisan leaders of World War II, after the war he was able to reform Yugoslavia as a Federal Republic. Although a communist, Tito's independent mindset, evidenced in part by the fundamental principles contained in the Yugoslavia's post-war constitution, led to a public split from Stalin in the spring of 1948. This event, coupled with Tito's "unambiguous commitment to the Yugoslav idea embodied in the omnipresent slogan 'Brotherhood and Unity' (*bratstvo i jedinstvo*)," led to the country's emergence as an independent, non-aligned, socialist state, which held together for nearly 50 years.[8]

Yugoslavia was a remarkably liberal and open society for a communist state.[9] That such a diverse country held together for so long was mainly due to its leader's personality, many believed. A practical and pragmatic ruler, Tito balanced freedoms for its constituent republics and, later, ethnic minorities with interventions, sometimes repressive, by the federal government when deemed necessary for stability. One example of the pragmatic choices that Tito made to help keep the federation together involved the Serbians, the most powerful region of Yugoslavia and a group that could threaten his power. With "a weak Serbia means a strong Yugoslavia" an unspoken rule of Titoism, it was the only one of the six republics to be internally sub-divided, with the autonomous provinces of Kosovo in the south and Vojvodina in the north.[10]

On the whole, while even "the most refined manipulation could not put feuding among Yugoslav nationalities to rest. Prior to the crisis … at the end of the 1980s ... ethnic rivalry was a muted theme in national life. … Yugoslavia's commitment to the ethic of *bratstvo i jedinstvo* was profound, and appreciated by a clear majority of citizens."[11] That the republic held together at all, let alone for a decade, after Tito's death in May 1980, a turbulent and difficult decade for Yugoslavia, may be a tribute to the strength of his grand experiment. The nation even hosted the 1984 Winter Olympics in Sarajevo.[12] Nevertheless, in the end, the experiment failed.

Yugoslavia's decline during the 1980s was brought about partly by economics, and partly by the inherent difficulties of dealing with a rapidly changing bipolar Cold War world without its founding leader, or at least a suitable successor, at the federal level. As time passed, and destabilizing influences strengthened, others

8 Ibid., 71.

9 Payam Akhavan and Robert Howse eds., *Yugoslavia the Former and the Future: Reflections by Scholars from the Region* (Washington, DC: Brookings Institution, 1995), 7. Others warn against confusing liberalism with a degree of tolerance.

10 Nation, *War in the Balkans*, 72–3; Peter Lyddon, *The Fall and Rise of Coercive Diplomacy in the Balkans*, monograph (Fort Leavenworth, KS: School of Advanced Military Studies, 1996), 20.

11 Nation, *War in the Balkans*, 73–4.

12 Karl Mueller, "The Demise of Yugoslavia and the Destruction of Bosnia: Strategic Causes, Effects, and Responses," in *Deliberate Force: A Case Study in Effective Air Campaigning: Final Report of the Air University Balkans Air Campaign Study*, ed. Robert C. Owen (Maxwell AFB, AL: Air University Press, 2000), 5.

came to view the federation as fatally flawed. Its final collapse was enabled by the rise of nationalist leaders within its constituent states who used these two factors, and ethnic identity, to their political advantage.

By the middle of the decade "the preconditions of a revolutionary situation were apparent."[13] In addition to high unemployment and inflation, personal savings were largely depleted. Polemics between the republics, where the politicians were asserting their rights to sovereignty, and the federal level, where the focus continued to be on economic reform, were becoming increasingly polarized. Politicians began openly resorting to themes of nationalism in their debates, and they weren't alone.[14]

The final, rapid disintegration of the federation, in just two and one-half years, occurred in three distinct stages Dušan Janjić believes. The first was the break-up of the single party system (1989) while the second was the introduction of a multi-party system through local elections (1990–1991). The final period, which began with the start of two civil wars in Slovenia and Croatia in June 1991 and ended with the outbreak of war in Bosnia-Herzegovina in April 1992, thrust the disintegrated Yugoslavia's into "an interethnic conflict between Croats, Muslims, and Serb."[15]

While Slovenia was the first battleground of the emerging civil war, it was a stage-managed affair from the very beginning, lasting only 10 days and ending with less than 70 killed.[16] Croatia was not to be so fortunate and there the fighting continued for six months. The world community quickly took notice and attempted to manage the conflict, with the U.S. calling for Europe to take the lead in what most in America considered to be a European problem.

Here the political aspirations of the EC would be put to a test. In the "hour of Europe," EC mediators were quickly involved with the belligerent parties, trying to bring hostilities to a close—with largely poor results.[17] From July to the end of that year, the EC "brokered no less than fourteen ceasefire agreements, all of which were violated virtually before the ink was dry."[18] Frustration on the continent grew to a point that editorial comments like "pathetic" and "debacle" were appearing in major European papers.

---

13 Susan Woodward, *Balkan Tragedy: Chaos and Dissolution After the Cold War* (Washington, DC: Brookings Institute, 1995), 73.

14 Ibid. She catalogs the escalation of these intra-state tensions, 82–113.

15 Dušan Janjić, "Resurgence of Ethnic Conflict in Yugoslavia: The Demise of Communism and the Rise of the 'New Elites' of Nationalism," in *Yugoslavia the Former and the Future: Reflections by Scholars from the Region*, ed. Payam Akhavan and Robert Howse (Washington, DC: Brookings Institution, 1995), 30.

16 Nation, *War in the Balkans*, 108. He reports 65 killed, 187 wounded; Woodward, *Balkan Tragedy*, 146.

17 The term "hour of Europe," was coined by Luxembourg Foreign Minister and EC official Jacques Poos. Nation, *War in the Balkans*, 120.

18 Ibid., 121.

On 25 September, the UN finally became involved largely because of the inability of the EC to resolve the problem. Approving UNSCR 713, a resolution calling for "a general and complete embargo on all deliveries of weapons and military equipment to Yugoslavia," was the first step in a three-month concerted effort that would result in the fifteenth, and final, ceasefire in January 1992.[19] In February, UNSCR 743 established the UN Protection Force (UNPROFOR) that would eventually include 14,000 multi-national troops protecting four areas inside of Croatia.[20] The six months of fighting ended with tens of thousands dead and hundreds of thousands of refugees and displaced persons, not to mention significant physical damage to towns and villages.[21]

Two critical insights emerged from the civil war in Croatia that would continue to haunt conflict management efforts in region in the future. First, it was clear that the EC, the UN, and individual nations were, eventually, willing to get involved with diplomatic efforts in the Balkans region. But it was also clear that there was great disagreement between governments with regards to how important the Balkans region actually was and, perhaps more critically, how much "blood and treasure" should be expended in solving the region's problems.

Woodward notes, "the instruments available to the EC ... in their attempts to influence the Yugoslav conflicts were primarily economic incentives or penalties and moral suasion."[22] In effect, there was no military "stick" to go along with the diplomatic "carrot." Opinions on the use of force ranged from ambivalence at best, to outright rejection. Even the UN mandate, a classic peacekeeping effort, assumed "an in place ceasefire, consent of the warring parties, neutrality between former belligerents, and limit[ed] rules of engagement ... to case of self defense."[23]

The second insight gained from this short conflict was that by utilizing such a disjointed, uncoordinated approach without the means to enforce any agreements, those seeking to manage the conflict ceded the initiative to the belligerents. Without the threat of lethal force, or at least the low likelihood of such an event, the opposing forces could continue their actions subject only to negative public opinion. As a result, the conflict ended only when the belligerents realized they were heading towards military stalemate and that it was then in their best interests to cease hostilities.

The third of the civil wars in the former Yugoslavia was by far the most devastating. As Woodward notes in the introduction to her book *Balkan Tragedy: Chaos and Dissolution After the Cold War*, the human cost from the six-month war in Croatia

19 Ibid., 124; Woodward, *Balkan Tragedy*, 180, 401. UNSCR 713 was the first of 67 resolutions passed in roughly the next three and a half years.

20 Woodward, *Balkan Tragedy*, 402; Nation, *War in the Balkans*, 124–6. The four non-contiguous areas were eastern and western Slavonia and northern and southern Krajina.

21 Woodward, *Balkan Tragedy*, 146; Nation, *War in the Balkans*, 119.

22 Woodward, *Balkan Tragedy*, 175.

23 Nation, *War in the Balkans*, 125.

> soon looked mild in comparison with the war in neighboring Bosnia-Herzegovina. In the first two years alone, the death toll in Bosnia was greater than 70,000. More than 2 million people of a population of 4.3 million became refugees or were displaced to other parts of the republic, expelled from their homes—probably never to return—by fear, war, or nationalist extremists aiming to rid their town or village of people whose ethnicity was different.[24]

But even calling this war a "civil war," which is what its outward appearances seemed to indicate, masked a complexity that helped contribute to the difficulties experienced by those trying to manage the conflict.

Some observers viewed this war as one of Serbian "aggression and expansionism," a position generally adopted by the U.S., while others viewed it as an internal struggle, a true civil war, a position adopted by the British and French among others. If it were truly the former, then armed intervention on behalf of Bosnia-Herzegovina would seem to be the required action. If the latter, then peace-keeping or peace-enforcement would be appropriate.[25] From this confused starting point, complicated mediation efforts involving the U.S., the Europeans, and the Russians would eventually emerge.

The fighting in Bosnia began in April 1992 and raged until the October 1995 ceasefire. The major belligerents were the Bosnian Serbs, supported by Slobodan Milošević in Belgrade, the Bosnian Croats, supported by Franjo Tudjman in Zagreb, and the government of Bosnia-Herzegovina led by Alija Izetbegović.[26] At the beginning of the war, Serb-dominated territories were predominantly in the northern, near three of the Serb UNPROFOR-protected areas just inside the border in Croatia, and the southern portions of the country. The Croats and Bosniaks, on the other hand, were strongest in the center, the Croats towards the west, the Bosniaks to the east.[27]

The specific operational details of the various offensives that made places such as Mostar, Srebrenica, Banja Luka, and Tuzla recognizable names on world media broadcasts have been recorded elsewhere and are outside the specific scope of this study.[28] Suffice to say that in areas such as these, and many others, one ethnic faction sought to gain final control of an area by killing, or forcibly expelling, other factions from their homes, farms, and businesses, a continuation of the ethnic

---

24 Woodward, *Balkan Tragedy*, 1–2. Nation records similar figures: Nation, *War in the Balkans*, 125.

25 See, for example, Mueller, *Demise of Yugoslavia*, 12–13; Woodward, *Balkan Tragedy*, 7, 147.

26 Nation, *War in the Balkans*, 152.

27 The 1991 census determined that 44 percent of Bosnia-Herzegovina was Muslim, 31 percent Serb, 17 percent Croat, and 5 percent "other." Nation, *War in the Balkans*, 149. An excellent map of Bosnia-Herzegovina outlining the country's ethnic composition by district can be found at fig. 8-1.

28 See, for example, Nation, *War in the Balkans*, 149–222, among others.

cleansing seen earlier in Croatia.[29] What is important for this study, however, is how those who were attempting to stop the violence tried to do so, and what role precision-guided munitions may have played in that effort.

The path from inaction to Operation DELIBERATE FORCE during the civil wars of Yugoslavia was, as Mueller correctly observes, "one of gradual but inconstant escalation."[30] As previously noted, there was little consensus amongst nations and international organizations about what should be done, and what, if anything, could be done to enforce those decisions. Just as the UN was establishing the four enclaves under UN protection in Croatia, its attention turned Bosnia.

The international community's first reactions were largely expressive. In a series of statements and resolutions, the UN Security Council variously "appealed," "noted with deep concern," and "demanded" in response to the rising violence. As with the arms embargo begun in September 1991, and expanded to include NATO ships in July 1992, these "resolutions ... sanctions ... peacekeepers where there was no peace to keep, celebrity visits to embattled Sarajevo, empty threats, and endless mediation ... produced considerable sound and fury but did little to deter the dynamic of conflict on the ground."[31] The drift towards direct military action, which might have been able to stop the suffering, was agonizingly slow.[32]

In October 1992, UNSCR 781 established a "ban on military flights in the airspace of Bosnia-Herzegovina."[33] NATO set up Operation SKY MONITOR which recorded more than 500 violations in the next six months. While David Owen writes that this operation initially worried Bosnian Serb leader Radovan Karadžić, his angst soon passed as it became clear that NATO had no authority to enforce the ban.[34] In April 1993, UNSCR 816 finally authorized "Member States to take all necessary measures to ensure compliance with the ban on flights."[35]

---

29 One observer, Tadeusz Mazowiecki, noted that "ethnic cleansing is not a consequence of this war, but rather its goal." Quoted in Nation, *War in the Balkans*, 158.

30 Mueller, *Demise of Yugoslavia*, 17.

31 Nation, *War in the Balkans*, 171.

32 Even after the change in U.S. Administrations, Madeleine Albright would note that "with a new President, a wary Secretary of State [Warren Christopher], a negative Pentagon, nervous allies, and crises in Somalia, then Rwanda and Haiti blowing up, we weren't prepared to run the risks of leadership in Bosnia." Madeleine K. [Korbel] Albright with Bill Woodward, *Madam Secretary: A Memoir* (New York: Mirimax Books, 2003), 181.

33 Woodward, *Balkan Tragedy*, 407; Mueller, *Demise of Yugoslavia*, 19; Nation, *War in the Balkans*, 171.

34 Lyddon, *Fall and Rise of Coercive Diplomacy*, 31.

35 Woodward, *Balkan Tragedy*, 412; Mueller, *Demise of Yugoslavia*, 19; Nation, *War in the Balkans*, 171.

Between April and June 1993, the UN approved a series of resolutions extending the UNPROFOR mandate to include six "safe areas" within Bosnia.[36] The last of these resolutions, UNSCR 836, while again keeping very tight restrictions on the ground troop's ability to use lethal force, included provisions for NATO aircraft to fly close air support (CAS) missions in support of ground troops. The DENY FLIGHT mandate was expanded to "include the possibility of launching non-CAS air strikes to deter or retaliate for attacks against peacekeeping forces in Bosnia."[37]

While these final three measures finally provided some "teeth" to the efforts to end the war in Bosnia, in practice the authorization to use military force was difficult to obtain, rarely used, and generally ineffective. As U.S. Deputy Secretary of State Strobe Talbott noted at a November 1995 National Press Club luncheon, "the mechanism for approving air strikes was too cumbersome to be effective and until 1995, only provided 'pinprick air strikes.'"[38] As events transpired, the first combat action of DENY FLIGHT occurred on 28 February 1994, when four Yugoslav Super Galeb attack aircraft were shot down by USAF F-16s. The first CAS mission was launched the next month.[39]

That same month, a significant political development took place which fundamentally altered the political-military landscape in Bosnia. Following three months of negotiations under U.S. auspices, the Bosnian Croats and Muslim Bosniaks signed the "Washington Framework Agreement," ending their hostilities and allowing their newly combined forces to focus on the Bosnian Serbs.[40] Following the end of a four-month ceasefire negotiated by former U.S. president Jimmy Carter, fighting resumed between the Bosnian-Croat alliance and the Bosnian Serbs.

This time, however, under intense international pressure to end the seemingly endless conflict, NATO began to use air power in a much more effective fashion. At the same time, concerted U.S.-led parallel diplomatic efforts sought to bring all sides to the negotiating table. Frustrated by the continued defiance shown by Milošević and the Serbs, on 27 August U.S. negotiator Richard Holbrook increased pressure by publicly threatening a "six- to 12-month campaign of air strikes

36 Woodward, *Balkan Tragedy*, 307.

37 Mueller, *Demise of Yugoslavia*, 19; Lyddon, *Fall and Rise of Coercive Diplomacy*, 31; Woodward, *Balkan Tragedy*, 307.

38 Quoted in Lyddon, *Fall and Rise of Coercive Diplomacy*, 31–2. Details of his remarks are contained in National Press Club *Record*, Volume XLV, No. 42. 16 November 1995. A short explanation of this complex process can be found in Bradley S. Davis, "The Planning Background," in *Deliberate Force: A Case Study in Effective Air Campaigning: Final Report of the Air University Balkans Air Campaign Study*, ed. Robert C. Owen (Maxwell AFB, AL: Air University Press, 2000), 38–47.

39 Mueller, *Demise of Yugoslavia*, 21. That mission did not drop any ordnance.

40 Woodward, *Balkan Tragedy*, 314; Nation, *War in the Balkans*, 183; Mueller, *Demise of Yugoslavia*, 20.

against the Bosnian Serbs 'to level the playing field' in the conflict."[41] When a mortar attack on a Sarajevo marketplace the very next day was definitively tied to the Serbs, Operation DELIBERATE FORCE was underway.[42]

Operation DELIBERATE FORCE (30 August–14 September 1995): The Air War in Bosnia

That the two-week air campaign, known as Operation DELIBERATE FORCE, was ever carried out successfully is much more of a surprise than it might first seem. It involved an incredibly complex system of coordination, notification, and negotiation between two international organizations, and a number of their member states. As Bradley Davis explains, "NATO military forces fulfilled United Nations Security Council resolutions (UNSCR) and NATO political mandates by assisting UN political efforts to bring peace to the region of the former Yugoslavia."[43] It was, at the time, a unique multinational operation involving two distinct chains-of-command.

DELIBERATE FORCE: Command Structure

On the UN side, the Secretary-General, Boutros Boutros-Ghali, was in overall charge of the body's political and military efforts. He assigned Yasushi Akashi as his special representative for the country, where he headed the political and military efforts of the United Nations Peace Forces (UNPF) and UNPROFOR from Zagreb.[44] The military head of the UNPF was French Lieutenant General Bernard Janvier, while the head of UNPROFOR was British Lieutenant General Rupert Smith.

On the NATO side, the organization's Secretary-General was Willy Claes. Day-to-day political activities were handled by the North Atlantic Council (NAC), while the Military Committee (MC), the supreme military authority, provided military guidance to major NATO commanders including the Supreme Allied Command Europe (SACEUR). The local NATO commander was Admiral Leighton Smith, commander-in-chief of Allied Forces Southern Europe (CINCSOUTH).[45]

---

41 Mueller, *Demise of Yugoslavia*, 28.

42 Nation, *War in the Balkans*, 191; Mueller, *Demise of Yugoslavia*, 28–9.

43 Davis, "The Planning Background," 37.

44 The terms UNPF and UNPROFOR can be confusing. The UN military forces in Yugoslavia were designated UNPROFOR from February 1992 until the spring of 1995. At that time, the overall military command became known as UNPF, which had three subordinate commands. The troops in Bosnia then became known as UNPROFOR. The other two commands were in Croatia and Macedonia.

45 Davis, "Planning Background," 38–47; Christopher M. Campbell, "The Deliberate Force Air Campaign Plan," in *Deliberate Force: A Case Study in Effective Air*

As evidenced by Strobe Talbott's comments highlighted above, the dual command and control structures had proved less than optimal during DENY FLIGHT. Indeed, there were cases where the delays in notification meant that legitimate targets, under watch by NATO aircraft, were able to slip away under cover before permission to strike was received.[46] One example of the inherent inefficiencies of this bifurcated structure is that, until he delegated the authority to Akashi in February 1994, the UN Secretary-General himself was the only one who could approve UNPROFOR requests for NATO air support.

By the summer of 1995, the UN and NATO had cobbled together a workable arrangement between the two organizations. While, as Davis points out, these "connections would not have lasted forever ... [and] the operational effectiveness of the planning process was lower than it could have," enough compromise and cooperation existed for the operation to succeed. First, the NATO Combined Air Operations Center (CAOC) and the UN Air Operations Control Center (AOCC) were talking. Next, General Smith, the UNPROFOR commander and USAF Lieutenant General Michael Ryan, in charge of NATO's air operations, were coordinating targets. Finally, the UNPF commander, Lieutenant General Janvier, and CINCSOUTH, Admiral Smith had, with the blessings of their political masters, completed a "dual key" agreement whereby the two officers jointly held the release authority for air strikes.[47]

DELIBERATE FORCE: Campaign Planning

A core part of what eventually would become the DELIBERATE FORCE air campaign plan was CINCSOUTH Operations Plan (OPLAN) 40101. This plan, originally developed in 1993 in response to the expanded response options made possible by UNSCR 836, was routinely updated but had never been designed as a full blown air campaign.[48] Recognizing this, in April 1995 General Ryan, now physically at the CAOC in Vicenza, Italy, independently began a "close hold" effort to develop an appropriate campaign plan that would cover full-scale operations over Bosnia. As target nominations came from his planning staff in Naples, he also reached back to the Air Staff's Checkmate office in Washington for assistance. The campaign plan was further refined in May and early June then briefed to Admiral Smith later that month.[49]

Campaigning: Final Report of the Air University Balkans Air Campaign Study, ed. Robert C. Owen (Maxwell AFB, AL: Air University Press, 2000), 90–93.

46 Davis, "Planning Background," 42.

47 Ibid., 47–57. Between February 1994 and the London Summit in July 1995, Akashi held the UN key. Ryan C. Hendrickson, *Diplomacy and War at NATO: The Secretary General and Military Action after the Cold War* (Columbia, MO: University of Missouri Press, 2006), 78.

48 Campbell, "Deliberate Force Air Campaign Plan," 94–7.

49 Davis, "Planning Background," 55–6.

CINCSOUTH in turn took the plan to the NATO/UNPF Joint Targeting Board, on 14 August, for coordination and to begin the approval process. While the proposal began to work its way through UN channels, Admiral Smith presented the plan up the chain-of-command on the NATO side. Both organizations approved the plan which contained three sets of targeting categories which effectively created three options for execution.

Lieutenant General Ryan was delegated authority for final target-list selections, a task which he never delegated further down the chain-of command due to the negative political impact that would occur if an error was made.[50] With the plan now approved, General Ryan's planners worked to flesh out the final target list, which centered on four key target categories: integrated air defense systems (IADS); command, control, and communications (C3); lines of communication (LOCs); and ammunition storage sites.[51] In a very real sense, the final plan reflected much of the thinking of Colonel John Warden and his Checkmate/Black Hole brethren. The final sequencing work was completed on 22 August 1995, just a week before it was executed.[52]

DELIBERATE FORCE: Execution

The Sarajevo marketplace bombing on 28 August put into motion Operation DELIBERATE FORCE. Admiral Smith turned the NATO "key" that day, while General Smith turned the UN "key" the next day following confirmation of Serb responsibility for the attack.[53] The air strikes began early on the morning of 30 August and continued throughout the day. As with DESERT STORM, heavy emphasis was placed on taking out key components of the IADS and the destruction of large ammunition storage facilities. Two days later, on 1 September, the UN "turned off" its key.[54]

While not unexpected, it is perhaps understandable that such a move was not appreciated by all involved in the operation, particularly for those well acquainted

50 Ibid., 57. Ryan's decision was to cause some delays when NATO wished to attack a time-sensitive "pop-up" target. Campbell, "Deliberate Force Air Campaign Plan," 111.

51 Other sources outline five or six adding, variously, "fielded forces," by which is meant heavy weapons not the people who man them, and "leadership" target categories. Campbell, "Deliberate Force Air Campaign Plan," 108–09.

52 Davis, "Planning Background," 58. The IADS work built upon an earlier NATO plan, Operation DEADEYE. Campbell, "Deliberate Force Air Campaign Plan," 101.

53 Campbell, "Deliberate Force Air Campaign Plan," 110. General Janvier was in France attending his son's wedding, which is why General Smith made the initial decision. Mark J. Conversino, "Executing Deliberate Force, 30 August–14 September 1995," in *Deliberate Force: A Case Study in Effective Air Campaigning: Final Report of the Air University Balkans Air Campaign Study*, ed. Robert C. Owen (Maxwell AFB, AL: Air University Press, 2000), 131; Hendrickson, *Diplomacy and War*, 78.

54 Campbell, "Deliberate Force Air Campaign Plan," 114; Hendrickson, *Diplomacy and War*, 79.

with earlier "graduated response" efforts such as Vietnam.[55] The pause did, however, allow for the intense diplomatic efforts to continue, although when the Serbs did not comply with UN/U.S. demands the bombing resumed on 5 September.[56] Once strikes resumed, and despite some bad weather, the operation continued to enjoy considerable success, so much so that General Ryan began to worry about running out of the targets that had been approved on 14 August. By 14 September, the two "key holders," Admiral Smith and General Janvier agreed that the military objectives were largely met and that another pause was needed.[57]

During this two week period, the combined force flew more than 3,500 sorties. U.S. forces, Air Force, Navy, Army, and Marine Corps, flew more than 2,000 of these missions including 1,499 of which can be considered "strike" sorties.[58] These missions dropped a total of 572 LGBs, and launched 13 TLAM and 10 SLAM missiles against targets in Bosnia.[59] The campaign enjoyed considerable success, destroying nearly 87 percent of the targets sets that were developed before the war.

In one example of the devastating effects of these strikes, 19 of 20 DEADEYE targets were destroyed, shutting down the air defense network as desired. At a greater level of detail, damage to 290 of 357 (81 percent) of individual aimpoints met or exceeded desired levels.[60] In a final, telling statistic, the "hit effectiveness," a measure of how many bombs were needed to meet damage requirements for a given DMPI, i.e. specific aim point, was one-half of that in the Gulf War just four years previous.[61]

DELIBERATE FORCE: Results and Outcomes

The fact that the bombing campaign resumed on 5 September has been viewed by some as "the single most important decision made during the entire course of DELIBERATE FORCE." Ambassador Holbrooke has observed that "the resumption of the operation was the 'most critical moment of the bombing' and

---

55 Conversino, "Executing Deliberate Force," 139.

56 Mark C. McLaughlin, "Assessing the Effectiveness of Deliberate Force: Harnessing the Political-Military Connection," in *Deliberate Force: A Case Study in Effective Air Campaigning: Final Report of the Air University Balkans Air Campaign Study*, ed. Robert C. Owen (Maxwell AFB, AL: Air University Press, 2000), 191–2.

57 Campbell, "Deliberate Force Air Campaign Plan," 114–15; Hendrickson, *Diplomacy and War*, 79–80.

58 Richard L. Sargent, "Deliberate Force Combat Air Assessment," in *Deliberate Force: A Case Study in Effective Air Campaigning: Final Report of the Air University Balkans Air Campaign Study*, ed. Robert C. Owen (Maxwell AFB, AL: Air University Press, 2000), 335.

59 Sargent, "Combat Air Assessment," 338–9; Hendrickson, *Diplomacy and War*, 83–85.

60 Sargent, "Combat Air Assessment," 343.

61 Ibid., 347.

that 'if the bombing had not resumed that day, the negotiations would have been very adversely affected.'"[62] This seems high praise from a man who had previously written that the former Yugoslavia was "the greatest collective security failure of the West since the 1930s."[63]

By the time the second ceasefire was called on 14 September, the will of the Bosnian Serbs had been dulled. "The Bosnian Serb Army (BSA) ceased hostilities, complied with requirements to withdraw heavy weapons from total-exclusion zones, and began negotiations for a peaceful settlement."[64] Only one NATO aircraft, a French *Mirage 2000K* was lost to Serbian fire.[65] Most importantly, all parties were brought to the negotiating table. On 5 October, President Clinton announced a 60-day ceasefire and U.S.-supervised talks were held in Dayton, Ohio 1–21 November 1995. The resulting Dayton Peace Accord was formally signed in Paris on 14 December.[66]

The Dayton Accord was, as Nation points out, "an agenda for peace, not a finished architecture."[67] Operation JOINT ENDEAVOR, conducted under NATO command, brought a multi-national Implementation Force (IFOR) into Bosnia-Herzegovina to enforce the end of hostilities and to provide a safe and secure environment to permit the peace process to move forward. After one year, the IFOR evolved into the Stabilization Force (SFOR), a force designed to deter hostilities and allow Bosnia to move forward with its rebuilding process.[68]

Unfortunately, while Dayton began the rebuilding process in Bosnia-Herzegovina, it did nothing to address the fragility in other regions of the Balkans. The fragmentation and intrastate tensions that had begun with Slovenia and Croatia, and erupted in Bosnia, would continue in Macedonia, Montenegro, and, most importantly, Kosovo. Within three years, NATO would once again be involved in combat operations in the former Yugoslavia.

The situation as it developed in Kosovo towards the end of the 1990s had actually been set in motion in 1989 when the province's autonomous status was rescinded. With that action, the Kosovar Albanian majority was effectively subjected to a Serbian occupation. The situation was "untenable, and the observation that in Yugoslavia 'everything started with Kosovo and everything will finish in Kosovo' quickly became commonplace."[69]

62 Quoted in Campbell, "Deliberate Force Air Campaign Plan," 114.

63 Richard Holbrooke, *To End a War* (New York: Random House, 1998), 21.

64 Sargent, "Combat Air Assessment," 340.

65 The crew was captured and held by the Serbians.

66 Nation, *War in the Balkans*, 192–3.

67 Ibid., 194.

68 Ibid., 195-200. SFOR operations ended in 2004. The State Department reports that by 2005, half of the 2 million displaced persons had returned to their former homes, and 93 percent of the property seized or occupied has been returned to its rightful owner. Fact Sheet: "Ten Years of Dayton Progress," available at: www.state.gov/r/pa/prs/ps/2005/57195.htm (accessed 4 December 2008).

69 Nation, *War in the Balkans*, 223.

There was much "history" between the Serbs and the Albanians, with the latter fighting against the former in both world wars. Kosovo had been incorporated into Yugoslavia after World War II, but Tito had had to use a heavy hand in the area for more than two decades afterwards. When that pressure was eased, protests against the federal government soon began. He responded with a policy of "liberalization," which eventually resulted in the province's autonomous status in the 1974 constitution.[70]

In 1990, the Democratic League of Kosovo (LDK) began pushing for national independence. Subsequently, in September, the Kosovo Assembly declared independence, but was immediately dissolved by the Serbians. The Assembly moved underground, again declaring independence one year later. In 1992, as the former Socialist Republic of Yugoslavia was fragmenting, Serbia and Montenegro formed the Federal Republic of Yugoslavia, although this "nation" was not recognized until 2000. In 1993 the so-called Kosovo Liberation Army (KLA) emerged from the underground nationalist movement as a quasi-military force for active resistance, beginning a low level campaign of violence against the Serbian authorities by early 1996.

The KLA campaign remained at the "nuisance" level until early 1998. To this point, U.S. diplomats had been courting Milošević, hoping he would help move the implementation of the Dayton Accords forward.[71] When Serbian authorities decided to suppress this hitherto small group in January, the crackdown backfired. The ensuing violence, particularly in early March, was seen as too heavy-handed. It swelled the ranks of the KLA, and quickly turned Western opinion against the Serbs.[72] Mindful of the limitations of the Dayton Accord, and not willing to repeat the mistakes made in the long run-up to DELIBERATE FORCE, American diplomats soon began concerted efforts to keep the violence contained.[73]

On March 5, the NAC expressed concern and promised to engage to prevent escalation. The International Contact Group, active during the Bosnian crisis, also returned to the forefront. On March 9 the group "condemned 'the use of excessive force by Serbian police against civilians' as well as 'terrorist actions by the Kosovo Liberation Army." The UN became involved at the end of March, passing UNSCR 1160 (31 March 1998) which condemned the "excessive use of force," froze foreign assets and placed an arms embargo on the FRY.[74] But just as was the case earlier in Bosnia, there was no effective coercive tool in place. As a result, negotiations during the spring and summer designed to stop the Serb offensive were largely fruitless.

Tensions between the nations seeking to end violence in Kosovo were high. As NATO sought to find ways to influence the Serbs, for example, Russia flatly

70 Ibid., 223–4.

71 Nation, *War in the Balkans*, 228.

72 Ibid., 225–9.

73 Holbrooke, *To End a War*, 357.

74 Nation, *War in the Balkans*, 234–5; Hendrickson, *Diplomacy and War*, 100–101.

refused to sanction any resort to air strikes under either UN or European auspices. NATO, although internally divided, argued that it could act without new UN mandates, relying rather on existing UN resolutions. On 24 September 1998, NATO increased pressure on the Serbs by issuing "Activation Warnings" for two military response options, the *"Limited Air Response,"* quick punitive strikes, and a *"Phased Air Campaign,"* a multi-phase air campaign.[75]

A week later, on 1 October, pressure was again increased as NATO issued "Activation Requests" for both options, concurrent with a request that NATO begin "Activation Orders" (ACTORDS). Intense negotiations between Ambassador Holbrooke, Slobodan Milošević, and others, and the very real possibility of a NATO air attack, resulted in the 13 October announcement of an "acceptance in principle" of NATO's terms by Serbian President Milan Milutinović. Despite this diplomatic capitulation, NATO issued the ACTORDS anyway, then immediately enacted a 96-hour pause to allow Belgrade to demonstrate compliance. Significantly, while NATO suspended the planned air operations on 27 October, they did not cancel the ACTORDS, instead "reserving the right to execute them at a later date if necessary."[76]

One critical flaw in the negotiations that had just been completed was that they had not included the Kosovar Albanians. As might have been expected, as Serbian troops withdrew from some areas, the KLA moved troops back in, and the two sides were soon fighting again. A report of a massacre of civilians by Serbian troops in Račak, on 15 January 1999, galvanized Western opinion. In only weeks, NATO forces would, once again, be engaged in air strikes in the Balkans.[77]

By the end of the month, the International Contact Group was trying to organize talks between the FRY and the Kosovar Albanians in a French châteaux, Rambouillet, near Paris. Simultaneously, the NAC delegated authority for approving air strikes to NATO Secretary-General Javier Solana.[78] Negotiations between the two warring parties went on through February and into the middle of March without resolution. While the Kosovar Albanians finally signed the accord on 19 March, the Serbs would not. On 23 March, Solana ordered General Wesley Clark, SACEUR, to commence operations. Operation ALLIED FORCE was underway.[79]

75 Nation, *War in the Balkans*, 237–8.

76 Ibid. Albright was not happy with the progress made. "We were acting 'like gerbils on a wheel,' in constant motion but getting nowhere." Albright and Woodward, *Madam Secretary*, 392.

77 Western diplomats did not want to be accused of downplaying atrocities again. See James Gow, *Triumph of the Lack of Will: International Diplomacy and the Yugoslav War* (New York: Columbia University Press, 1997).

78 Nation, *War in the Balkans*, 240–42.

79 Hendrickson, *Diplomacy and War*, 104.

Operation ALLIED FORCE (24 March–10 June 1999): The Air War in Kosovo

The decision to launch Operation ALLIED FORCE was, from the start, a controversial one. As opposed to DELIBERATE FORCE, with its dual UN/NATO mandate and chain-of-command, this operation was fully a NATO affair. While some in Europe still wanted UN Security Council endorsement, the poor performance of UNPROFOR and the awkwardness of the "dual key" arrangement made it easier to keep the UN out of the military aspects of the operation.[80] This has called into question, some believe, its underlying legal authority.[81] In fact, NATO, rather than trying to find consensus when it needed to do something about Kosovo, merely let each member of the alliance determine a legal justification of its own for the use of force.[82]

Others, perhaps more cynical, claim that, despite NATO rhetoric about the situation in Kosovo being "a humanitarian disaster of unprecedented dimensions," only around 2,000 people had died and 250,000 had been internally displaced.[83] Tragic certainly, but less of a disaster than witnessed even in our earlier case studies, and nothing on the order of, say, Rwanda. The real reason for the war, these critics claim, was simply to punish the Serbs.[84] While true that there was an anti-Serb bias in many of the nations involved, this seems too simplistic an answer. A humanitarian crisis was underway and it needed to be addressed.

ALLIED FORCE: Command Structure

Since the operation was entirely a NATO affair, the command structure was much more streamlined. As described in Ryan Hendrickson's *Diplomacy and War at NATO: The Secretary General and Military Action after the Cold War*, Secretary General Javier Solana played a much more active role during ALLIED FORCE than Claes had in DELIBERATE FORCE.[85] The military officer in charge of

80 Ibid., 95.

81 See Katariina Simonen, *Operation Allied Force: A Case of Humanitarian Intervention*, Athena Papers (Garmisch-Partenkirchen: Partnership for Peace Consortium of Defense Academies and Security Studies Institutes, 2004). Lambeth argues that existing UNSCRs were sufficient. Lambeth, *NATO's Air War*, 14n.15.

82 For a summary of alliance member's justifications regarding intervention see Thomas F. Walsh III, *Operation Allied Force: Setting a New Precedent for Humanitarian Intervention?* Thesis (Monterey, CA: Naval Post Graduate School, 2000), 11–52.

83 Nation, *War in the Balkans*, 244.

84 Ibid., 244–5. Nation records that, in *Winning Ugly: NATO's War to Save Kosovo* (Washington, DC: Brookings Institution Press, 2000), Daalder and O'Hanlon go so far as to declare that Rambouillet itself was not meant to come to a diplomatic solution, but "to create a consensus in Washington and among NATO allies that force would have to be used."

85 Hendrickson, *Diplomacy and War*, 110–13.

ALLIED FORCE was U.S. Army General Wesley Clark, wearing his SACEUR "hat." To provide the senior leadership necessary to synchronize U.S. and NATO operations, Admiral James Ellis, Jr., the officer who was now CINCSOUTH, was appointed Commander, Joint Task Force Anvil. The air campaign was to be run by Lieutenant General Michael Short, USAF, who now was Commander, Allied Air Forces Southern Europe, the position previously held by General Ryan during DELIBERATE FORCE, although in practice General Clark remained very much involved in General Short's day-to-day operation.[86]

ALLIED FORCE: Campaign Planning

The NATO planning effort that underpinned the "*Limited Air Response*" and "*Phased Air Campaign*" first discussed in September 1998 "rested upon an existing NATO planning effort that had been underway since June ... [which had] no less than 40 air campaign variants."[87] The ground option, although initially developed, was shelved almost from the very beginning as neither the U.S. or NATO wished to send soldiers into Kosovo.[88] This total reliance on air power and missiles as a means of coercion was readily apparent from the start.[89] Though perhaps shortsighted, it is understandable because, as Michael Clarke points out, "Airpower is frequently regarded as embodying the ultimate instrument of military coercion."[90] Simply put, "precision air strikes left the politicians in control; airpower could strike without affecting much else; it could send political 'messages' to opposing leaders; and an air campaign could be called off and the force dispersed at any time."[91]

Planners were under considerable constraint throughout and the prevailing belief was that a short series of punitive strikes would convince the Serbs to capitulate. As a result, a fully integrated plan along the lines of DESERT STORM or DELIBERATE FORCE was never developed. In fact, writing later, General Wesley Clark would candidly admit that "in the case of Kosovo, there simply was no detailed planning."[92]

---

86 For an explanation of command and control arrangements, see Department of Defense, *Kosovo/Operation Allied Force After Action Report: Report to Congress* (31 January 2000), 16–21.

87 Nation, *War in the Balkans*, 237; Lambeth, *NATO's Air War*, 11.

88 Clark was particularly enthusiastic about using ground troops. Solana tempered that excitement in the interests of Alliance cohesion. Hendrickson, *Diplomacy and War*, 111–12. Clark acknowledges this point. Wesley K. Clark, *Waging Modern War: Bosnia, Kosovo, and the Future of Conflict* (New York: Public Affairs, 2001), 166.

89 Nation, *War in the Balkans*, 246.

90 Quoted in Dag Henriksen, *NATO's Gamble: Combining Diplomacy and Airpower in the Kosovo Crisis 1998–1999* (Annapolis: Naval Institute Press, 2007), x.

91 Henriksen, *NATO's Gamble*, x.

92 Clark, *Waging Modern War*, 439.

What was developed instead was a three-phase campaign. Phase I included antiaircraft defense and command posts, Phase II added strategic infrastructure targets below the 44th parallel, while in Phase III the capital Belgrade would be attacked. Believing the conventional wisdom that a short, punitive effort would be sufficient to bring the Serbs to the table, the final plan envisioned striking only 50 pre-approved targets in a span of two to four days.[93] This effort was, as events unfolded, wholly inadequate and ad hoc planning began almost as soon as the war began.

The follow-on process of adaptation has been described as "chaotic, unscripted, and confused."[94] While General Clark quickly outlined a shift from the strategy of intimidation to one designed to "systematically and progressively attack, disrupt, degrade, devastate ... and destroy" the Serb's ability to fight, his focus remained primarily on the Serbs fielded forces.[95] This set up a deep philosophical, and sometimes heated disagreement between SACEUR and his subordinate air commander that, while at first glance seems eerily reminiscent to that of DESERT STORM, was to prove much more problematic for day-to-day operations.[96]

General Short, in keeping with DESERT STORM and DELIBERATE FORCE concepts of operation, wanted to attack "the head of the snake" by striking key government buildings in Belgrade and other strategic infrastructure targets.[97] While some accommodation was reached, serious tensions would remain. Starting in the first week of April, both approaches were taken simultaneously. Similar to the structure of the Iraqi air campaign, fielded forces would be degraded while strategic attacks sought to convince the Serbs to capitulate. To their credit, planners did not self-limit their target selections due to perceived political sensitivities. Instead, they selected targets of military significance and let the formal review process determine which, in a quickly changing campaign, could be struck.[98]

ALLIED FORCE: Execution

When the operation commenced, NATO had approximately 344 aircraft (214 U.S./130 Allied) at its disposal. On the opening night of the war, 55 cruise missiles began the assault followed by 120 fixed-wing aircraft attacks against some 40,

93 Nation, *War in the Balkans*, 247–8; Clark, *Waging Modern War*, 178; Lambeth, *NATO's Air War*, 13.

94 Nation, *War in the Balkans*, 248.

95 Quoted in Nation, *War in the Balkans*, 248. This shift in strategy was formally approved by NATO one month into the campaign: Clark, *Waging Modern War*, 203; Lambeth, *NATO's Air War*, 23.

96 Lambeth, *NATO's Air War*, 190–92.

97 Nation, *War in the Balkans*, 252; Lambeth, *NATO's Air War*, xix-xx.

98 Lamb, *Golden Nuggets*, 13.

primarily air defense, Phase I targets.[99] The second night of the war saw heavier attacks against similar types of targets. In addition to these force protection type strikes, the remainder of the first week of strikes concentrated on fielded forces, although these were often difficult to find. NATO had hoped for a quick capitulation. Instead, the Serbs used this time to accelerate their ethnic cleansing.

As the realization dawned that this would not be a quick operation, additional aircraft were dedicated to the effort, which began to shift to a strategy that included more strategic targets. Phase II targets were authorized to be struck on Day 4, while, for all intents and purposes, the Phase III targets were attacked starting on Day 9.[100] On 3 April, the Yugoslav and Serbian Interior Ministries, in downtown Belgrade, were destroyed. On 21 April NATO brought the war directly to the political and media elite in Serbia who supported Milošević, when they attacked state radio and television stations owned by his wife, and the businesses of his family and friends.[101]

In a highly controversial attack, on 23 April NATO attacked the building that housed Serbian state television.[102] The attacks on critical infrastructure were placing considerable pressure on Serbian leaders, although the attacks in Serbian troops were much less successful. The civilian population however, with the notable exception of the few unfortunate casualties, was largely unaffected. As General Clark later observed, this was "the only air campaign in history in which lovers strolled down riverbanks in the gathering twilight and ate at outdoor cafes and watched the fireworks."[103]

In early May all that changed. In an effort to place more pressure on Serbian leaders, NATO began to bring the war directly to the civilian population. First, the electrical power grid was temporarily degraded. Factories were also attacked, which caused massive unemployment. During the last two weeks of the war more permanent damage was done to the electrical power grid of the country, although with greater care than in DESERT STORM, "leaving millions without electricity and water service."[104] Unfortunately, as in Iraq, there were noteworthy errors made.

On 12 April a railroad bridge near Leskovac was attacked by an American aircraft while a train was approaching. Unfortunately, the first missile that was

99 Lambeth, *NATO's Air War*, 22. The first night cruise missiles included USAF CALCM plus U.S. and UK TLAM. This marked the first non-U.S. use of TLAM in combat.

100 By June more than 1,000 aircraft would be involved. Lambeth, *NATO's Air War*, 25–9; Nation, *War in the Balkans*, 251.

101 Lambeth, *NATO's Air War*, 39; Department of Defense, *Kosovo/Operation Allied Force After Action Report*, 7–8. Julian Tolbert explores the concept of "crony attack" in Julian H. Tolbert, *Crony Attack: Strategic Attack's Silver Bullet?* Thesis (Maxwell AFB, AL: Air University Press, 2006).

102 NATO insisted the latter was a legitimate target because the Serbs used it to disseminate war propaganda. Nation, *War in the Balkans*, 253.

103 Quoted in Lambeth, *NATO's Air War*, 43.

104 Ibid.

launched failed to drop its targeted span so the train did not stop. Not realizing this, the crewmember released the second missile, which hit the train. Reportedly, 10 people were killed and many more injured.[105] Two days later, civilian vehicles intermixed with a military convoy were struck. While, as with the *Al Firdos* bunker, these would arguably fall under the double-effect provision of the just war tradition, the press reports were not positive for NATO.[106]

The incident with the greatest political ramifications, however, occurred on 7 May. Due to an error in validating targets, the Chinese Embassy in Belgrade was attacked. Three were killed and 20 were wounded. While the U.S. immediately apologized, President Clinton personally wrote to Chinese President Jiang Zemin, the Chinese still believed the attack was intentional.[107]

As the war lengthened, the option of a ground campaign was again broached. While discussed, the option gained little traction. UN and NATO ground and logistics planners were prohibited from conducting proper campaign planning, although options were discussed. These ran into considerable logistical and political problems. Simply put, Kosovo was a hard place to get to and logistically support, while those surrounding nations whose support would be necessary had populations who were either solidly opposed to NATO's actions or had governments who had significant concerns about the parties involved.[108]

While combat operations were underway, numerous attempts at finding a diplomatic solution were launched. Interestingly, the track that was to be ultimately successful originated with the Russians. Before the war, Russian Prime Minister Evgenii Primakov had been involved with finding a way to avoid conflict, but had taken a hard anti-Western stance once combat began. In April, Russian President Boris Yeltsin appointed a seasoned diplomat, former Prime Minister Victor Chernomyrdin, as a special peace envoy. Chernomyrdin softened Primakov's earlier stance and worked to increase Russian influence and find a compromise acceptable to all parties.

In late May, he traveled to Belgrade to meet with Milošević, who agreed to a peace proposal on 3 June. On 10 June the peace proposal and ceasefire agreements were codified into UNSCR 1244, which formally ended the war.[109] During the 78-day campaign, NATO pilots flew just over 38,000 sorties, about 10,500 of which were strike missions, and dropped more than 28,000 munitions.[110] The conflict also witnessed the combat debut of the B-2 stealth bomber and the first use of a

105 Clark, *Waging Modern War*, 254; Lambeth, *NATO's Air War*, xvii.

106 Clark, *Waging Modern War*, 254–5.

107 Nation, *War in the Balkans*, 253; Clark, *Waging Modern War*, 296–7; Lambeth, *NATO's Air War*, 144–7.

108 Lambeth, *NATO's Air War*, 45–6; Department of Defense, *Kosovo/Operation Allied Force After Action Report*, 8.

109 Nation, *War in the Balkans*, 256–9; Clark, *Waging Modern War*, 373.

110 Lambeth, *NATO's Air War*, 61; Others report 14,000. Nation, *War in the Balkans*, 251; Lamb, *Golden Nuggets*, 1.

new type of PGM, the Joint Direct Attack Munition (JDAM). The JDAM, with its onboard GPS guidance system could be launched effectively in situations, such as heavy cloud cover, where LGBs could not be used.[111] Only two alliance aircraft were lost.

ALLIED FORCE: Results and Outcomes

In the immediate aftermath of the operation, Western politicians were quick to claim unconditional victory. But, as Nation correctly points out, "the Chernomyrdin-Ahtisaari agenda was significantly different from that which had inspired the original Rambouillet accord," the Serbians had won some concessions.[112] In the first place, although the war had been NATO's, the post-war process was now firmly back under UN control.

This effectively meant that the soon to be deployed Kosovo Peacekeeping Force (KFOR) and all refugee return operations, designed to be NATO-led under Rambouillet, were now UN or joint operations. The Chernomyrdin-Ahtisaari agreement also restricted the NATO force's freedom of movement. While Rambouillet would have permitted NATO troops to travel throughout the FRY, their movements were now restricted to Kosovo.[113] While NATO won the war, the Serbs acceding to the Alliance's demands, many of the underlying tensions remained. After a year, only about a third of the 200,000 Serbs who had lived in Kosovo before the air campaign had returned. Kosovar Albanians also reversed the ethnic cleansing process in the region making "a mockery of allied intentions to recreate an authentically multicultural environment."[114]

Foreign troops, although in much smaller numbers, today still deploy in a peacekeeping role. The province was run by a UN organization, the UN Interim Administration Mission in Kosovo (UNMIK), for almost ten years. In February 2008, after years of negotiations as to its final status, Kosovo declared its independence from Serbia. In June, the UN shut down UNMIK and the new Republic of Kosovo continues to move forward to ensure equitable stability within its borders.

An Analysis of the Performance of Precision-Guided Munitions

The public perception of the near-perfect performance of PGMs in these two case studies only served to enhance the reputation that had first emerged during Operation DESERT STORM. While it is true that in DELIBERATE FORCE almost all of the weapons dropped by American Forces were PGMs, the percentage

111 Lambeth, *NATO's Air War*, 61.

112 Nation, *War in the Balkans*, 261.

113 Ibid.

114 Nation, *War in the Balkans*, 263; Clark, *Waging Modern War*, 412.

in ALLIED FORCE is probably much lower than most observers realize. But the fact that collateral damage and civilian casualties were kept very low solidified the perception of perfection and likely raised expectations for future employment. One passage about the operation in Bosnia by Paul Gillespie captured the excitement about PGMs:

> Operation DELIBERATE FORCE ... avoided ground fighting altogether by relying upon a relatively abundant supply of precision guided munitions. Post campaign analysts concluded that this approach was successful both because it did not 'kill people and destroy property to an extent that would cause world opinion to rise against and terminate the operation,' and because it gave the Serb faction minimal opportunity to fight back and inflict casualties on NATO and United Nations forces. In other words, near-exclusive reliance upon PGMs allowed military power to be applied in an antiseptic way that never risked the loss of public support, either at home or abroad.[115]

As with Operation DESERT STORM, because this case study occurred in the relatively recent past, there is unfortunately not as much unclassified effectiveness data as might be hoped for. However, there is enough publicly available data to draw meaningful conclusions.

Part I: The Weapons and Their Employment

Returning to the first level of analysis in the analytical framework, the effectiveness and manner of use of precision-guided munitions in the Balkans will be examined. The weather, during both the Bosnia and Kosovo operations, was to play a significant factor in causing aircrew both to abort missions prior to weapons delivery and to lose the ability to spotlight the target during a weapon's time-of-flight thereby causing a "miss."[116] This factor, coupled with the aforementioned enhanced sensitivity to collateral damage and civilian casualties set the stage for potentially negative trends during these two operations. Fortunately, as events transpired, that was not the case.

Operational effectiveness At first glance, the overall effectiveness of U.S. PGMs in Operation DELIBERATE FORCE would appear to deliver a mixed verdict. As noted earlier, the number of targets destroyed, aimpoints damaged, and the significantly better "hit effectiveness" would all seem to be positive outcomes. However, against the strict requirement that a "hit" was an observed impact inside a 15-foot CEP, only about 60 percent of LGBs dropped achieved that requirement.[117] Clearly, while the 15-foot CEP is not an issue, the low

115 Gillespie, *Precision Guided Munitions*, 208–9.

116 Clark alludes to this in Clark, *Waging Modern War*, 274.

117 Sargent, "Combat Air Assessment," 338–9.

number of direct hits could be seen as a cause for concern. Delving a bit deeper begins to help mitigate this viewpoint.

The USAF conducted an evaluation of why weapons did not strike their intended aimpoints. Of the 123 LGBs that missed their target, 64 percent were attributed to reasons not related to the weapon itself such as: aircraft malfunctions; crew error; and weather-related issues.[118] Taking these factors into account, and if the statistics held true for all misses, it is possible that more than 80 percent of the weapons were functioning properly as they were delivered.

Finally, the available data does not indicate how close any of the misses were to their aimpoints. The high degree of target destruction would seem to indicate that at least some landed inside the target area, but outside the strict 15-foot criteria for success.[119] Additionally, the fact that General Ryan's concerns about a civilian atrocity caused by NATO bombs were, in fact, unrealized should also be viewed as a positive indicator regarding weapons accuracy.

The performance of TLAM and SLAM stand-off missiles is, unfortunately, not consistently positive. Thirteen TLAM were launched against targets in Bosnia. Depending on the unclassified source used, either nine or eleven hit their intended target.[120] If the latter number was accurate, it would match historical averages. SLAM, on the other hand, did not perform well in Bosnia with only two of ten hitting their assigned targets. Post-strike analysis did, however, uncover the prime cause of these failures. Seven of eight failures, it emerged, resulted from "command-guidance failure due to GBU-15, [the USAF LGB glide bomb,] data-link interference."[121] In other words, Navy aircrew could not control the weapons because the command data-link, which provides steering commands, was electronically "jammed."

A variety of PGMs were also used during Operation ALLIED FORCE. However, in contrast to DELIBERATE FORCE, far more unguided weapons were used, by both U.S. and non-U.S. aircraft during the operation. In fact, of the 23,315 U.S.-delivered munitions expended in the war, only 6,728 (29 percent) were PGMs. Of the remainder, "a full one-third were general purpose Mk 82 unguided bombs dropped by B-52s and B-1s during the war's final two weeks."[122] Additionally, reliable information regarding PGM effectiveness is harder to locate for Operation ALLIED FORCE, although that which is publicly available is largely positive.

118 Ibid.," 339.

119 General Ryan made this rule. Richard L. Sargent, "Deliberate Force Tactics," in *Deliberate Force: A Case Study in Effective Air Campaigning: Final Report of the Air University Balkans Air Campaign Study*, ed. Robert C. Owen (Maxwell AFB, AL: Air University Press, 2000), 305.

120 Sargent, "Combat Air Assessment," 399; "RGM/UGM-109B/C Tomahawk." *Jane's Naval Weapons Systems*, Issue 26, Apr 98, 2.

121 Sargent, "Combat Air Assessment," 339; Conversino, "Executing Deliberate Force," 153.

122 Lambeth, *NATO's Air War*, 64–6.

A total of 238 TLAMs were fired from U.S. submarines and surface ships and, for the first time, a UK Royal Navy submarine, the HMS *Splendid.* Of those fired by U.S. ships, 83 percent reached their assigned aimpoint, a figure in keeping with historical averages. Portions of an internal U.S. Navy document, obtained by *Jane's Defence Weekly*, which analyzed the effectiveness of TLAM during this operation found "no collateral damage and no evidence of any shoot-downs."[123]

Another PGM "success story" during ALLIED FORCE was that of JDAM. This new system was not actually a new weapon but rather a bolt-on kit designed to turn "dumb bombs" into GPS-guided PGMs. This innovation allowed inexpensive PGMs to be used in weather situations that normally precluded the use of LGBs, thereby requiring expensive weapons such as TLAM, and also permitted a single aircraft to drop a number of "launch and leave" weapons on a single pass. Because it relies on GPS, instead of a laser beam, the weapons publicly acknowledged CEP, at 13 meters (40 feet), is slightly larger than that of LGBs.[124]

The weapon was so new that initial production batches were being rushed into operational use, and eventually 656 of the new weapons were used in Kosovo.[125] The weapons did, however, functioned as designed demonstrated by the fact that, "[90] percent of JDAM struck within 40 feet of their intended targets."[126] Of note, with its advanced synthetic aperture radar, the B-2 aircrews were able to refine the final DMPI and achieved CEPs about half of that advertised, which brought it much closer to LGB-like accuracy.[127] As Gillespie, among others, notes, Lieutenant General Michael Short declared the B-2 and JDAM "the No. 1 success story" of ALLIED FORCE.

The success of CALCM launches during this operation is difficult to ascertain. One source reports that "During the first two weeks, no B-52 succeeded in launching all eight of its CALCMs. In one instance, six out of eight were said to have failed."[128] Some report more than 100 were fired. Similarly, SLAM, which did not fare well during DELIBERATE FORCE, was considered a "success story" in this operation, although details at the unclassified level are scarce.[129]

It is also quite difficult to track the precise overall performance of LGBs in this conflict. In one sense, that is probably good news. Once a new weapon has lost its media novelty, is used routinely, and operates within design specifications, it ceases to be newsworthy. As a means to gauge their performance, and to segue

---

123 Bryan Bender, "Tomahawk Achieves New Effects in Kosovo," *Jane's Defence Weekly*, 19 July 2000. Of the HMS *Splendid's* missiles, 17 of 20 (85 percent) hit their aimpoint.

124 Later JDAM variants added an LGB seeker.

125 Gillespie, *Precision Guided Munitions*, 210.

126 Ibid., 211.

127 Lambeth, *NATO's Air War*, 91.

128 Ibid., 20n.11.

129 Department of Defense, *Kosovo/Operation Allied Force After Action Report*, 93; The USAF study "Air War Over Serbia," has never been declassified.

into the next portion of the analysis, it should be noted that in the Department of Defense's *Kosovo/Operation Allied Force After-Action Report: Report to Congress*, the operation is called "the most precise and lowest collateral damage air operation in history."[130] Indeed, "one postwar estimate concluded that only 20 of 23,000 munitions dropped by NATO in the Kosovo campaign caused civilian casualties."[131] While admittedly more empirical data would be better, that which is available indicates that, across most categories, PGMs once again demonstrated a low CEP and a high rate of weapons functional effectiveness. As a result, the trend line here is to be seen as positive

Manner of employment During both the planning for and execution of Operation DELIBERATE FORCE, high emphasis was placed on minimizing human suffering, both civilian and military. Early in the planning process, each target was "linked to a degree of risk of collateral damage and to infrastructure targets that, if destroyed, would cause undue hardship on the civilian populace."[132] As Sargent notes, because of this keen focus "everything from rules of engagement to aircraft, weapons, and tactics selections was driven to some degree by concerns about collateral damage."[133] Aircraft dropping unguided weapons, for example, only did so in target areas with little chance of collateral damage.[134]

This attention to limiting human suffering was, in contrast to the electrical grid situation in DESERT STORM, clearly transmitted down the chain-of-command. For example, as the planning entered its final stages in August, General Ryan "instructed his AIRSOUTH strategy cell and the CAOC operational planners to limit collateral damage as much as possible ... and to ensure the protection of NATO forces to the highest degree."[135] For the general, this meant "no civilian casualties or undue military casualties."[136] The goal of the air campaign was to destroy the equipment that was important to the Serbs, thereby driving them to the negotiating table, not to physically decimate their forces. The policy has to be seen

130 Department of Defense, *Kosovo/Operation Allied Force After Action Report*, xiv.

131 Gillespie, *Precision Guided Munitions*, 230. Others cite 28,000. See Lambeth, *NATO's Air War*, xx.

132 Richard L. Sargent, "Deliberate Force Targeting," in *Deliberate Force: A Case Study in Effective Air Campaigning: Final Report of the Air University Balkans Air Campaign Study*, ed. Robert C. Owen (Maxwell AFB, AL: Air University Press, 2000), 283.

133 Ibid., 287.

134 Dutch Air Force F-16s, using unguided bombs, had some of the most accurate hits "in the history of air warfare." Michael O. Beale, *Bombs Over Bosnia: The Role of Airpower in Bosnia-Herzegovina*, Thesis (Maxwell AFB, AL: Air University Press, 1997), 37.

135 Davis, "Planning Background," 58.

136 Ibid.; Conversino, "Executing Deliberate Force," 159–60; Beale, *Bombs Over Bosnia*, 37.

as a success as Milošević himself later admitted that only 25 Serbs were killed during the air campaign.[137]

Another overt manifestation of this policy can be seen in the decision by AIRSOUTH planners not to target bridges to help avoid inflicting civilian casualties. While this policy was later loosened when it was feared that Serbs were going to reinforce around Sarajevo, General Ryan did place restrictions on pilots attacking bridges, some of which placed them in greater danger themselves, which not all appreciated even though they understood the political realities of the region.[138] In addition, the Rules of Engagement (ROE) under which the pilots flew all strike missions required that the pilot positively identify not just the target that had been assigned, but the exact aimpoint (DMPI) that had been selected.[139] Because this was not always possible due to weather, missions were sometimes aborted.

As in Bosnia, exceptional efforts were made to limit both friendly losses and collateral damage on the ground in Kosovo, efforts which were, to a great extent, successful. To enable the former, aircraft were mandated to generally fly at altitudes of 15,000 feet or above, although this was relaxed to a degree as the war progressed. As in DESERT STORM, while this helped keep allied pilots out of harm's way, the Serbian IADS was never fully suppressed, it complicated their target identification efforts. General Clark's planners were even required to estimate the potential collateral damage, and unintended civilian casualties, a process which continued once hostilities began and caused problems for those planning in theater.[140]

Through the first three weeks of the campaign, Clark records, "We had struck almost 1,000 targets, and there had been only eight incidents of serious civilian losses. I had to stand up for the airmen, who were flying the most precise and error-free campaign ever conducted, so far as we could see."[141] But concerns about civilian casualties never diminished and, as the campaign entered its final stages, Clark was under increasing pressure from NATO nations to reduce them to zero.[142] Interestingly, harkening back to both Vietnam and DESERT STORM, when the likelihood of collateral damage was very low, and the tactical payoff high, area coverage weapons such as cluster bombs and area bombing by B-1s and B-52s were permitted.[143]

---

137 Conversino, "Executing Deliberate Force," 161; Beale, *Bombs Over Bosnia*, 37.

138 Conversino, "Executing Deliberate Force," 150, 160; Sargent, "Deliberate Force Targeting," 293–4.

139 Campbell, "Deliberate Force Air Campaign Plan," 114.

140 Clark, *Waging Modern War*, 179, 225.

141 Ibid., 297.

142 Clark, *Waging Modern War*, 339; Lambeth, *NATO's Air War*, 27.

143 Clark, *Waging Modern War*, 336–7. CBUs were approved for use early in the war, and then withdrawn after a couple of civilian casualty incidents. Later, during the

The ROE restrictions, designed in part due to NATO's heightened sensitivity to collateral damage, and seen as "uncompromisingly restrictive" by many, also included stringent "no hit" target categories.[144] Additionally, where required by potential collateral damage concerns, some aircraft were launched with 500-pound LGBs when 2000-pound weapons would have been the norm.[145] While there were a few notable exceptions, these measures were quite successful. Michael Ignatieff describes one attack on a military communications tower, concluding "the mangled frame lies on its side across a water-filled bomb crater, located just 25 feet from the homes of the Albanian families."[146]

During the campaign only about 500 civilians were killed by the allied bombing effort, with various sources reporting 20, 30, or 90 incidents of collateral damage.[147] It was noted that the "intensive use of precision strike systems ... was critical in this regard."[148] When such unfortunate incidents did occur, changes in tactics often followed. In early May, for example, the use of CBUs was discontinued, while later that month bombing of bridges was shifted from day to night—in response to civilian casualties.[149]

Precisely because of such careful planning and execution, more evidence of an important but potentially complicating trend is also to be found during ALLIED FORCE. Just as with the unfortunate, but morally justifiable, Al Firdos bunker incident, the relatively few collateral damage cases garnered much greater attention than would have occurred in past conflicts. It seems clear that, along with the capability to strike precisely, there has emerged an expectation on the part of senior leaders and civilians alike that it will be done so in every case. As Thomas notes, "NATO's desire to avoid collateral damage was, in effect, in constant competition with both its goal of coercing Slobodan Milošević and its strong interest in incurring as few casualties among its own airmen as possible."[150]

Unfortunately, for many political and civilian observers, much, much better, in terms of reducing unnecessary human suffering is increasingly seen as insufficient. Admiral Ellis commented after ALLIED FORCE, "As a cautionary note, given the severe restriction on collateral damage, it was noted that precision guided is no

---

battle for Mount Pastrik, SECDEF approval for their use was once again given. The B-1s and B-52s began flying these missions in early May. Lambeth, *NATO's Air War*, 49.

144 Nation, *War in the Balkans*, 247.

145 Lambeth, *NATO's Air War*, 22.

146 Michael Ignatieff, *Virtual War: Kosovo and Beyond* (New York: Metropolitan Books, 2000), 91.

147 The government of Yugoslavia made varying claims of between 1,200 and 5,700 civilian deaths. Human Rights Watch places the number at between 488 and 527. Chris Finn, "The Broader Implications of the Increasing Use of Precision Weapons," *Air Power Review* 4.1 (Spring 2001): 38.

148 Nation, *War in the Balkans*, 251; Lambeth, *NATO's Air War*, xx, 219n.3.

149 Thomas, *Ethics of Destruction*, 163.

150 Thomas, *Ethics of Destruction*, 164.

longer 'good enough.'"[151] Ignatieff agrees saying, "the Western public has come to think of war like laser surgery. Displays of this kind of lethal precision at first awakened awe; now they are expected."[152] General Anthony Zinni, USMC, then-CINCCENT, worried that "the success of airpower has set the bar too high for future operations. 'The expectations are so great now: zero ... casualties, perfect execution, completely flawless."[153] Technology, created by humans is still fallible, as are those who employ it. While these concerns will be returned to in the concluding chapter, against the framework's two lines of inquiry in this portion of the framework, the trends lines are clearly positive.

Part II: The Question of Unintended Consequences

As with the previous case studies, the two operations in the Balkans provide valuable insight into the question of unintended consequences across all three lines of inquiry. It is worth noting, however, that this case study also highlights the interconnectedness of the individual portions of the framework. For example, while the slaughter of Muslims at Srebrenica in Bosnia, or of Albanians in Cuska, Podjevo, and Velika Krusa and elsewhere during Kosovo could reasonably be addressed in this section, they will instead be addressed in the final portion.

Rapid capitulation with unresolved tensions It is apparent from the narrative presented that, while the U.S. and European nations were certainly unenthusiastic about intervening in the quagmire that was the former Socialist Federal Republic of Yugoslavia, the humanitarian crisis would not evaporate of its own accord and would have to be dealt with eventually. When those interventions transpired, however, the reliance on an aerial assault utilizing PGMs did little to resolve underlying tensions, although they did serve to bring the parties to the negotiating table. In Bosnia, the UN and NATO waited an overly long time before waging a short campaign, with mixed results.

As Nation later wrote, "At Dayton the United States and its allies succeeded in imposing a fragile truce upon the warring factions, but failed to construct a viable framework for peace building. Eight years after the end of hostilities Bosnia-Herzegovina remains occupied by a NATO-led peacekeeping force, with the goal of a self-sustaining peace still out of reach."[154]

In Kosovo, on the other hand, a relatively short wait ended with a longer than anticipated campaign. As before, though, the underlying political and social issues were largely unresolved. As earlier noted, today there are still foreign troops on the ground in Kosovo helping that nation move forward into the future.

151 Quoted in Lamb, *Golden Nuggets*, 19.

152 Ignatieff, *Virtual War*, 92.

153 Quoted in John A. Tirpak, "Kosovo Retrospective," *Air Force Magazine* 83.4 (April 2000): 31.

154 Nation, *War in the Balkans*, 203–4.

Adverse asymmetrical responses There is also evidence to suggest that a reliance on precision-enabled air power, to police the region prior to the execution of Operation DELIBERATE FORCE, did result in asymmetric responses that caused the UN and NATO significant difficulties and loss of credibility in the eyes of the world. The issue of "blue hat" hostages is a prime example. On 10 April 1994, NATO aircraft bombed an enemy position for the first time in history following a Serb mortar attack that killed an UNPROFOR soldier. In response, General Ratko Mladic had his troops surround 150 UNPROFOR troops in the area.[155]

In another instance, on 24 May 1995, in an attack designed to punish the Bosnian Serbs for a violation of the Sarajevo safe area, NATO aircraft struck ammunition dumps near the city of Pale. In a move reminiscent of Saddam Hussein, the Serbs responded "by taking 370 UN peacekeepers hostage and chaining some of them to potential targets, thereby paralyzing the intervention."[156] Both cases highlighted the limitations of using air power when vulnerable forces are deployed in the region.

During Operation ALLIED FORCE, Milošević was again unable to directly counter the U.S./Allied air assault. However, he did employ a number of indirect, asymmetric means to attempt to disrupt coalition unity and play on the alliance's key concerns about collateral damage and a humanitarian crisis. These tactics included: "the use of terror tactics against Kosovar civilians; attempts to exploit the premium the alliance placed on minimizing civilian casualties and collateral damage; creation of enormous refugee flows to create a humanitarian crisis; and, the conduct of disinformation and propaganda campaigns."[157]

Long term suffering William Arkin, who wrote earlier of DESERT STORM, also wrote about Operation ALLIED FORCE. What he records is almost universally positive with regards to minimizing collateral damage and long-term suffering, observing that "NATO demonstrated an almost obsessive concern for civilian casualties and damage. ... Enormous efforts were made, largely successful, to minimize short- and long-term civilian effects. When mistakes did occur ... changes ... were implemented almost immediately."[158]

He catalogs that, despite Yugoslavian propaganda to the contrary:

---

155 Beale, *Bombs Over Bosnia*, 25.

156 Robert C. Owen, "Summary," in *Deliberate Force: A Case Study in Effective Air Campaigning: Final Report of the Air University Balkans Air Campaign Study*, ed. Robert C. Owen (Maxwell AFB, AL: Air University Press, 2000), 477; Beale, *Bombs Over Bosnia*, 29; Albright and Woodward, *Madam Secretary*, 185.

157 Lamb, *Golden Nuggets*, 2–3.

158 William Arkin, "Operation Allied Force: 'The Most Precise Application of Air Power in History'," in *War Over Kosovo: Politics and Strategy in a Global Age*, ed. Andrew J. Bacevich and Eliot A. Cohen (New York: Columbia University Press, 2001), 28.

> NATO did not bomb water installations, dams, or pharmaceutical plants, and there were no attacks on food storage or other targets immediately affecting the survival of the civilian population. Electrical-transformer stations and power lines were attacked ... [but] electrical generation remained off-limits in an attempt to minimize the long-term effect of bombing on the civilian population.[159]

The issue of electrical power was a particularly sensitive one during operations in Yugoslavia. The large degree of post-war human suffering in Iraq caused by the destruction of the electrical power system, in violation of the JFACC's original intentions, came to be seen "as a serious black mark against the way the war [DESERT STORM] was fought."[160] There was clear effort not to have a repetition of that outcome.

Not all are so positive however. Nation notes, "The decision to attack national infrastructure ... was without question an unusual one. Its impact on the Serbian people ... and upon the long-term economic well-being of Serbia and the entire Balkan subregion, was profoundly negative."[161] Thomas agrees when he writes:

> it must be acknowledged that even very precise bombing can exact a terrible toll on a civilian population. Casualties aside, NATO bombing set Yugoslav society back many years, inflicting an estimated $30 billion in damage to the economy, driving the unemployment rate toward 50 percent, and forcing down annual per capita income to less than $1000.[162]

Such realities promoted Gary Dempsey of the CATO Institute to pen an article, eerily reminiscent of some commentary regarding Vietnam, titled "Destroying Serbia in Order to Save It."[163] Similarly, Dr. Michael Evans observes that, "Before the air war, there were 45,000 refugees outside Kosovo; after the air war there were 855,000. Even if 80 percent return, there will still be 160,000 refugees—quadruple the number that existed before the air war. If this is victory, then it is a dark victory."[164]

Against the three metrics developed for this section of the framework, the trend lines must be seen as mixed for this case study. First, while it is clear that the use of PGMs as a catalyst to get parties to the negotiating table was, in both cases, ultimately successful, it is also clear that many of the underlying problems were not resolved by this action. The complicating factor is that solving those simmering problems was not the reason the bombing commenced. The goal was

---

159 Ibid., 24.

160 Thomas, *Ethics of Destruction*, 166.

161 Nation, *War in the Balkans*, 253.

162 Thomas, *Ethics of Destruction*, 178.

163 Gary Dempsey, "Destroying Serbia in Order to Save It," *CATO Institute* 8 June 1999: http://www.cato.org/pub_display.php?pub_id=5606 (accessed 17 December 2008).

164 Michael Evans, "Dark Victory," *Proceedings* 125.9 (September 1999): 37.

to stop the immediate humanitarian crisis, which was done. That political resolve, and the concrete steps needed to solve systemic problems in the region, were slower to follow cannot reasonably be attributed to the technology being studied. This trend line then, is flat.

On the other hand, for the second metric, that of evidence of a violent asymmetric response, the trend line is clearly negative. In the past, reliance on a more traditional ground-based force would have at least offered the possibility of a quick intervention when UN troops or civilians were threatened—although there is no way to be certain that such action would have been taken. Relying on precision air power alone, however, eliminated even that, perhaps remote, possibility of an intervention.

Finally, while the concerns of the last commentators above point to significant economic disruption to the Yugoslav people after the campaign, as Arkin points out the war did little long term damage to critical infrastructure. As opposed to the massive increase in infant mortality in Baghdad after DESERT STORM due to the destruction of electrical production capability, the allied effort was much more discriminate here. Additionally, in Bosnia and Kosovo, while it admittedly took longer than any party would have liked, Western intervention and ongoing commitment allowed a slow but steady rebuilding of the nations. The trend line then, while not steep, is positive.

Part III: The Question of Teleology

While the Vietnam case study was difficult to analyze for this portion of the framework, and DESERT STORM demonstrated a strong commitment on the part of senior decision-makers not to rely on the use of precision-enabled air power alone, here the exact opposite is true. The great reluctance on any nation's part to introduce ground combat troops into the theater, coupled with the general impotence of UNPROFOR and the strong desire to end each of the two conflicts entirely with no or low casualties using precision air power offers some interesting insights into the teleological analysis.

Reluctance to deploy conventional land forces In the summer of 1991, no party outside of Yugoslavia seemed willing to use military power, ground or air, in the region as part of any effort to stop the fighting in Slovenia and Croatia. Neither "the EC nor the CSCE [Conference on Security and Cooperation in Europe] had the military capability to act or even to mount a credible threat to intervene with force that might deter those willing to use force in Yugoslavia. The United States had publicly declared its unwillingness to use military power in the case and had insisted that NATO should not become involved anywhere 'out of area.'"[165]

In a series of statements in May 1992, the EC, CSCE, UN Security Council, and the U.S. State Department issued calls for an end to violence in Bosnia. The violence

165 Woodward, *Balkan Tragedy*, 165.

was seen on the television yet, paradoxically, "growing outrage over Serb conduct did not affect the international consensus opposing military intervention."[166] As fighting intensified in the summer of 1992, there was still a general reluctance to use military force in the region. Lyddon notes that "disagreements arose over how [to stop the fighting in Bosnia,] with the Europeans generally favoring a less active approach and the U.S. favoring the use of air strikes."[167]

Lawrence Eagleburger, former American Ambassador to Yugoslavia and later, for a short period of time, U.S. secretary of state, remarked in September 1992:

> I have said this 38,000 times, and I have to say this to the people of this country as well. This tragedy is not something that can be settled from the outside and it's about damned well time that everybody understood that. Until the Bosnians, Serbs and Croats decide to stop killing each other, there is nothing the outside world can do about it.[168]

In his memoirs of his time as U.S. secretary of state, Warren Christopher records that in February 1993, "We were all frustrated that we failed to stop the bloodshed, but we realized our options were limited. The Pentagon realized that air power alone could not prevail in the Balkan terrain, and no one had any enthusiasm for commitment of U.S. ground troops to force a settlement."[169] It would not be until mid-1995, that "the President decided that the United States should take a stronger hand, politically and militarily."[170]

In a series of resolutions between April and June 1993, the UN Security Council expanded the mandate for the UNPROFOR, to include six "safe areas" within Bosnia. Even then, however, the soldiers were only permitted "to use force in self-defense or in deterring attacks against the safe areas." They were not permitted to defend the safe areas against such attacks. Interestingly, the measure did authorize the use of "all necessary measures, through the use of air power, to support UNPROFOR."[171]

As this series of observations demonstrates, no nation was willing to commit to solving the humanitarian crisis using ground forces. As Susan Woodward points out, the longer the conflict went on without resolution, the greater the human tragedy became. Knowing that atrocities were taking place, yet not doing "much"

166 Nation, *War in the Balkans*, 171.

167 Lyddon, *Fall and Rise of Coercive Diplomacy*, 29.

168 Holbrooke, *To End a War*, 23.

169 Warren Christopher, *In the Stream of History: Shaping Foreign Policy for a New Era* (Stanford, CA: Stanford University Press, 1998), 345.

170 Christopher, *Stream of History*, 347.

171 Mueller, *Demise of Yugoslavia*, 15. Quotes from UNSCR 836 (June 4, 1993) reprinted in Woodward, *Balkan Tragedy*, 414.

because the region wasn't in "our national interests" raised disturbing questions of "strategic versus moral significance."[172]

The reluctance to deploy true combat forces in theater, and to instead rely on the odd NATO air strike to coerce the Serbs, was to have a significant impact on the Muslim population of Srebrenica. As Holbrooke records, "Emboldened by his successes in intimidating the U.N. peacekeepers, General [Ratko] Mladic now focused pressure on the three isolated Muslim enclaves in eastern Bosnia."[173] These three enclaves had been designated as "safe areas," and by the summer of 1995 all were filled with Muslim refugees. As noted in a Human Rights Watch/ Helsinki report, the UN was woefully under prepared to protect the safe areas as promised.

The UN had estimated it would need 34,000 troops to carry out the mission, however "only 3,500 were eventually deployed and only a few hundred of those were deployed in Srebrenica."[174] The UNPROFOR understood its mission was to "protect the civilian population of the designated safe areas against armed attacks and other hostile acts, through the presence of its troops and, if necessary, through the application of air power, in accordance with agreed procedure."[175] However, with far fewer troops than planned, and an unwieldy process to call in air support, the UN/NATO were ill prepared to stop this next round of ethnic cleansing.

On 6 July the Serb forces began to shell Srebrenica, entering the town on 10 July. During the next week, "the biggest single mass murder in Europe since World War II took place, while the outside world did nothing to stop the tragedy."[176] While exact details were not known at the time, Holbrooke acknowledges "there was no question that something truly horrible was going on."[177]

In fact, after days of indecision and denial of requests by General Janvier, the first NATO air strikes were launched after noon on 11 July. It was, as one UN official said, "a meager display. Too little, too late."[178] Later, the Red Cross would determine that just over 7,000 Muslims were slaughtered between 12 and 16 July, most dying in ambushes and mass executions.[179]

It seems readily apparent that both the U.S., in particular, and NATO, to perhaps a lesser degree, were equally reluctant to use ground combat troops in Kosovo. While early planning efforts had included a full range of ground operations, a

172 Woodward, *Balkan Tragedy*, 289–95.

173 Holbrooke, *To End a War*, 68–9.

174 Human Rights Watch/Helsinki, *Bosnia-Hercegovina: The Fall of Srebrenica and the Failure of U.N. Peacekeeping*, tech. rept. no. Vol. 7, No. 13 (New York: Human Rights Watch/Helsinki, 1995), 7.

175 Quoted in Human Rights Watch/Helsinki, *Bosnia-Hercegovina*, 7. Original in U.N. Security Council, "Report of the Secretary-General Pursuant to Resolution 844 (1993)," S/1994/555, May 9, 1994, 5.

176 Holbrooke, *To End a War*, 69.

177 Ibid.

178 Human Rights Watch/Helsinki, *Bosnia-Hercegovina*, 13.

179 Holbrooke, *To End a War*, 70. Elsewhere 8,000.

RAND study states "from mid-1998 onward, not only was this option [ground operations] shelved, no serious contingency planning for land operations was undertaken. The exclusive planning focus was on air and missiles strikes."[180] As if to eliminate any doubt, President Clinton declared "I do not intend to put our troops in Kosovo to fight a war."[181] Later, at a breakfast with General Wes Clark just prior to NATO's fiftieth anniversary summit, Secretary of Defense William Cohen remarked, "Nothing about ground forces. We have to make this air campaign work, or we'll both be writing our resumes."[182]

This publicly stated position was to have significant consequences for Albanians in the region. Some of the greatest massacres by Serbian forces, in Cuska, Podjevo, and Velika Krusa, occurred during the air campaign. According to one estimate made by the British government, about 10,000 ethnic Albanians were killed during ALLIED FORCE.[183] Also, within just the first few days of the air campaign thousands of Kosovar Albanians began fleeing their homes in what one author called "a ghastly exodus that would quickly reach near biblical proportions."[184] By the end of the 78-day campaign, the UNHCR would calculate that more than 848,000 people would flee the province, roughly 40 percent of the population.[185]

In addition to giving Milošević freedom of movement, and a freer hand in how he would seek to achieve his political objectives, definitively taking the ground option off the table extended the combat. Testifying before Congress after the war, Admiral Ellis stated that "ruling out the use of ground forces 'probably prolonged the air campaign.'"[186]

One positive observation to come out of this analysis is that this case may serve as a foil against Ellul's belief that human fascination with the image is an entirely negative trend. The images portrayed on televisions around the world, created a surge in calls for the various government to solve the humanitarian crisis. Within the Clinton administration, which was unhappy that public opinion was encouraging policies that it felt were not in the interest of the U.S., this was known as the "CNN curve." Testifying before Congress, Secretary of State Warren Christopher declared, "Television images cannot be the North Star of America's foreign policy."[187] But they were beginning to serve that purpose.

---

180 Quoted in Nation, *War in the Balkans*, 246. Originally in Bruce R. Nardulli et al., *Disjointed Warfare: Military Operations in Kosovo, 1991* (Santa Monica: RAND, 2002), 16.

181 Ibid.

182 Clark, *Waging Modern War*, 269.

183 Lambeth, *NATO's Air War*, 60.

184 Nation, *War in the Balkans*, 249; Lambeth, *NATO's Air War*, 26.

185 Nation, *War in the Balkans*, 249–50.

186 Quoted in Lamb, *Golden Nuggets*, 23.

187 Quoted in Woodward, *Balkan Tragedy*, 324.

Across the Atlantic, Douglas Hurd, the British Foreign Secretary in the summer of 1992, maintained that "the decision to commit troops had nothing to do with the media."[188] This observation seems hard to validate. As Hudson and Stainer firmly believe, "The catalyst which finally compelled the world to react was the combined force of the world's press and television. Probably never before had the influence of the media been more powerfully felt by governments of the world in the ensuing months and years of the war in Bosnia."[189]

Evidence has been presented in the narratives above to demonstrate that this trend was evident in the run-up to both Bosnia and Kosovo. What is fascinating is that, rather than focusing on the image to the exclusion of what lay behind it as Ellul intimates, here the image appears to have forced reluctant governments to act at the behest of an outraged populace. While such an observation does not fit neatly into the evaluative framework it is, nonetheless, enlightening.

Greater willingness to resort to lethal force While clearly reluctant to engage the warring parties with ground forces, it was also clear that over time the UN and NATO became increasingly willing, however reluctantly, to use air power in a much more deliberate and lethal fashion in Bosnia. The UN tended to be much more tentative in its willingness to use air power. In November 1994, in response to a Serbian attack against Croat and Muslim forces in western Bosnia, NATO prepared to strike the Serbian airfield at Ubdina. While the UN approved the strike, Secretary General Boutros Boutros-Ghali dictated that only the runways could be struck, the aircraft themselves were off limits. The runways were quickly repaired.[190] Later, air power became the enabler of DELIBERATE FORCE.

Similarly, while clear that the U.S. and NATO had no intention of deploying their ground troops into Kosovo, it is also clear that the willingness to use precision-enabled air power was aided by a striking misperception of how long the air campaign might last. In a televised broadcast on the evening of the first night of the war, Secretary of State Madeleine Albright declared to her U.S. audience, although the world was certainly watching too, "I don't see this as a long-term operation," words she regretted almost immediately.[191] It seems abundantly obvious that the willingness to wage what was expected to be a short air campaign *in the absence of clearly articulated political or military objectives* was mainly due to the perceived ability to use precision guided weapons and avoid incurring much negative political backlash. Despite these misperceptions, it is the willingness to use air power that is important here.

188 Warren Chin, "The Transformation of War in Europe 1945–2000," in *European Warfare 1815–2000*, ed. Jeremy Black (New York: Palgrave, 2002), 199.

189 Chin, "Transformation of War," 199.

190 Holbrooke, *To End a War*, 61.

191 Nation, *War in the Balkans*, 247; Lambeth, *NATO's Air War*, 19–20; Albright and Woodward, *Madam Secretary*, 408.

With all of this as supporting evidence, it is clear that greater suffering took place in both Bosnia and Kosovo due to the strong desire to forego the use ground forces and use air power as the sole military tool of choice, which opened up opportunities for the enemy. Thus, the trend line based on the first metric would initially be clearly negative for this case. The two evaluation metrics are, however, deeply interconnected.

The willingness to use precision air power alone, which directly led to combat operations and would be seen, at first glance, as a negative according to the second of the metrics, simultaneously led to the cessation of hostilities. Thus, what first appeared to be two negative trends actually becomes the first case where the later alleviation of suffering overcomes the initial analysis. Combined, the trend becomes only slightly negative as the suffering evidenced along the first line of inquiry is not fully offset by that which was alleviated along the second.

The three case studies examined so far have been, with the exception of DELIBERATE FORCE, extended combat operations. However, while two other long-term operations are currently underway, there are also shorter combat operations that rely heavily on PGMs, and sometimes last but a few minutes. Because the structure of these operations is significantly different, analysis of these strikes might offer new insights. As a result, the study will now turn to a series of contingency strikes as a final composite case study.

Chapter 7
Contingency Strikes: Expanded Political-Military Action Enabled By Precision

> Air power has become the instrument of choice in America's foreign policy.
>
> Major General Charles Wald, USAF

The three major case studies examined thus far have evaluated the impact of PGMs on human suffering in war as this particular technology has matured from being a promising new technology playing the role of a bit player towards the end of a much larger and longer conflict, to a key, but operationally limited, enabling component of a large combined operation, to, finally, the star player in two air power-only campaigns. Using the metrics developed in the evaluative framework, these case studies provide many useful insights that will be summarized in the concluding chapter.

At first glance, the next logical case studies to undertake would be the ongoing operation in Afghanistan and the recently completed war in Iraq. Unfortunately, most of the operational details needed for this effort such as weapons expenditures and effectiveness figures, remain necessarily classified, so information is quite limited. Additionally, the information that is available, however accurate it may prove to be in the future, is today really just "instant history."[1] These early narratives, observations, and recollections of a few key participants and other authors have not yet stood the test of time, research, and peer review. However, these operations will not be ignored, but reflected upon in a slightly different fashion the final chapter. There is another type of operation that, while perhaps seen by some as outside the traditional category of "war," fits neatly into this study's expanded definition and may offer insights that provide added support to, or fundamentally alter, the conclusions already being formed. This is the issue of "contingency strikes."

Contingency Operations

According to U.S. Department of Defense Joint Publication 1-02, a contingency is a "situation requiring military operations in response to natural disasters, terrorists, subversives, or as otherwise directed by appropriate authority to protect

1 Jamie Shea addresses this issue in "Instant History," *NATO Review* 49.2 (Summer 2001): 21–3.

U.S. interests."[2] Planning for contingency operations may occur over a long period of time, in anticipation of possible triggering events, or evolve quickly in response to a rapidly developing crisis. Such operations can be joint, those involving more than one U.S. service, combined, those involving more than one nation or, in rarer instances, operations that may involve forces from only a single American military service. Since the end of the Vietnam War, the U.S. has engaged in a number of contingency operations that are well known to the general public.

Operation EAGLE CLAW (April 1980), for example, was the ill-fated military operation to free the hostages taken from the American Embassy in Tehran. Operation URGENT FURY (October–December 1983) was the U.S.-led invasion of Grenada undertaken following the overthrow of the government of Prime Minister Maurice Bishop, while Operation JUST CAUSE (December 1989–January 1990) was the name for the American invasion of Panama. Finally, Operation RESTORE HOPE (December 1992–May 1993) was the U.S.-led operation to protect relief supplies being brought into Somalia.

Each of these operations clearly demonstrated a willingness on the part of the U.S. government to place American ground forces in harm's way in support of missions deemed within the national interests of the country. Increasingly, however, contingency operations in the form of short duration air strikes, thus outside of the more traditional model, have taken place. These contingency strikes, where no ground troops and relatively few Americans are placed at risk, have been enabled by the ready availability of precision-guided munitions in the form of aircraft-delivered munitions and air- and sea-launched cruise missiles. Seven of these strikes will be examined.

Operation EL DORADO CANYON (14–15 April 1986)

During the 1970s and 1980s, relations between the United States and Libya were tense, first over Colonel Muammar Qaddafi's support of international terrorism and later over his stated claim to the entire Gulf of Sidra, bounded by the ominously named "Line of Death." Following the election of President Ronald Reagan, the U.S. reasserted its right to sail in these international waters through "Freedom of Navigation (FON)" operations which began in August 1981. During the first FON operation, two Libyan SU-22 fighter aircraft closed on the American force, launched an air-to-air missile at a Navy F-14 fighter, and were subsequently shot down.[3] Later, in 1985, a series of terrorist incidents linked to the Libyan government resulted in the JCS commencing planning for a possible military response if further terrorist acts occurred.[4]

---

2 Joint Pub 1-02, s.v. "contingency."

3 John F. Lehman Jr., *Command of the Seas* (New York: Charles Scribner's Sons, 1988), 362.

4 Peter A. Huchthausen, *America's Splendid Little Wars: A Short History of U.S. Military Engagements: 1975–2000* (New York: Viking Press, 2003), 88–91; Todd R.

In April 1986 two incidents resulting in American deaths and injuries, a terrorist bombing of a Berlin discotheque and a bomb exploding on a TWA flight, placed in motion this retaliatory plan.[5] The American objective was to "send a grave signal to the world that it would not tolerate any more acts of terrorism."[6] In a developing pattern that would be repeated throughout any number of future contingency operations, the selection of targets was designed to "not only emphasize this message and convey the full measure of ferocity *but also keep civilian casualties near the targets low and minimize damage to U.S. forces.*"[7] To accomplish this, the JCS settled on a single pass, multi-axis nighttime air attack.

Navy carrier-based aircraft used unguided weapons to attack the Jamahariyah Barracks and Benina airfield, and also provided suppression of the Libyan air defenses. USAF F-111Fs, of the 48th Tactical Fighter Wing (TFW) and launched from Great Britain, attacked three target complexes, the Azziziyah Barracks and the terrorist training school at Sidsi Balal using LGBs, and the military side of Tripoli airfield using unguided weapons.[8] While surface-to-air missiles were launched, no Libyan fighters rose to meet the strikes. The results of the LGB drops appear, at first glance, to be mixed. As Werrell records,

> Results were less than hoped for, with only half of the 24 bombs dropped (of the 48 originally planned) hitting their targets. Four F-111s made successful bomb drops, six aborted, seven missed their targets, and one was lost. Bombs from two aircraft went awry, one hitting the French embassy and two others impacting 700 yards from their target, damaging two homes, killing 37 civilians, and injuring 130.[9]

But this account needs to be more closely examined to be properly understood, as it blends characteristics of the two types of attacks.

First, not all of the F-111s were carrying LGBs. In his summary, Werrell accounts for 18 aircraft. While this is indeed the total number of bombers in the USAF portion of the strike, only 12 of them were loaded with LGBs, accounting for the total of 48 bombs listed. Of that number, Daniel Bolger records six aircraft on target, one aircraft lost, one miss, and four aborts, which largely agrees with Werrell.

Phinney, *Airpower Versus Terrorism: Three Case Studies*, Thesis (Maxwell AFB, AL: School of Advanced Air and Space Studies, 2003), 12.

5 Joseph T. Stanik, *El Dorado Canyon: Reagan's Undeclared War with Qaddafi* (Annapolis: U.S. Naval Institute Press, 2003), 142–5; Lehman Jr., *Command*, 371.

6 Huchthausen, *Splendid Little Wars*, 93.

7 Ibid: emphasis mine.

8 Winnefeld and Johnson, *Joint Air Operations*, 84–7; Huchthausen, *Splendid Little Wars*, 94–5; Phinney, *Airpower Versus Terrorism*, 20.

9 Werrell, *Chasing the Silver Bullet*, 154.

However, Bolger also notes that five of the six aircraft that dropped a total of 60 unguided weapons were on target while one aborted.[10] Additionally, while the clear LGB "miss" was determined to be due to the crew error, one of the aborts occurred because another crew recognized the same situation and called off their attack in keeping with the assigned Rules of Engagement.

Looked at it in this perhaps more positive light, "Department of Defense spokesman Robert Sims' announce[ment] that 'only 1 to 2 percent of the bombs impacted on civilian areas.' Compared to the bombings of any previous war by any air force, this was a major accomplishment," seems accurate.[11] The overall strike was deemed a success as terrorist activity diminished for some time to come.[12] But the strike does serve as a reminder that even highly accurate weapons can go astray if aircrew proficiency and discipline are not maintained.

Operation PRAYING MANTIS (18 April 1988)

One part of the extended war between Iran and Iraq took place at sea, becoming known as the "Tankers Wars." In an effort to influence their respective adversary, or to punish those who were providing war loans, Iran and Iraqi each attacked merchant shipping in the Persian Gulf. Iraq attacked ships entering or leaving Iranian ports, while Iran attacked the shipping of those Gulf States that were providing economic support to Baghdad.[13] These operations began in 1984 and, as an example of how extensive these attacks became, in 1987 each belligerent attacked more than 85 vessels, an average of more than three attacks per week.[14] That year, the United States decided to take decisive action.

Concerned about possible interruptions to the world oil supply, and particularly anxious to avoid a repeat of the negative economic impact caused by earlier oil shortages, in July 1987 the Reagan Administration agreed to re-flag a number of Kuwait tankers and escort them into and out of the Persian Gulf through the Strait of Hormuz.[15] This operation, known as EARNEST WILL, would continue until September 1988. During this period, Iran was also suspected of seeding the shipping lanes of the Gulf with mines. In fact, the first U.S. re-flagged Kuwaiti vessel, the *Bridgeton* (formerly the *al-Rekkah*), struck a mine on the first EARNEST WILL

---

10 Quoted in Phinney, *Airpower Versus Terrorism*, 20. Part of the discrepancy can be attributed to the exact definition of what is a miss. While true that only four LGBs and twelve "dumb" bombs were direct hits, most others were close enough to do significant damage to their targets, without causing undue suffering. See Stanik, *El Dorado Canyon*, 184–93.

11 Blackwelder, *Long Road*, 22.

12 Werrell, *Chasing the Silver Bullet*, 154; Huchthausen, *Splendid Little Wars*, 93.

13 Michael A. Palmer, *On Course to Desert Storm: The United States Navy and the Persian Gulf*, Contributions to Naval History, vol. 5 (Washington, DC: Naval Historical Center, 1992), 109.

14 Ibid., table 9, 121.

15 U.S. law prevented the Navy from escorting foreign-flagged ships.

escort mission. In September of that year, U.S. forces finally gained definitive proof, catching the crew of the *Iran Ajr* in the act of laying mines. That vessel was subsequently attacked. On 14 April 1988, the American frigate *Samuel B. Roberts* struck a mine while patrolling the Gulf.[16] The U.S. was quick to respond.

Four days later, on 18 April, the U.S. Navy launched a one-day retaliatory effort known as Operation PRAYING MANTIS. The plan was to destroy two Iranian oil platforms and an Iranian frigate in a measured response to the mining of the American frigate.[17] As events unfolded, the day's effort was so successful that it "led to the destruction of about half of [Iran's] operational ships."[18] The two platforms, the Sassan and Sirri gas-oil separation platforms (GOSP), were chosen because they had been used by the Iranians to coordinate actions against shipping. Similarly, the Iranian frigate *Sabalan* was chosen as the desired surface target because it had attacked numerous surface merchant vessels and had been observed "concentrate[ing] [weapon's] fire on the crew quarters of tankers and other vessels."[19]

Operations began at 0800 that morning. While surface ships used their guns to destroy the two platforms, other surface ships and aircraft searched for the Iranian frigate.[20] While the *Sabalan* initially retreated and hid among some tankers, other Iranian vessels were more aggressive. As the day progressed, these vessels, the Iranian frigate *Sahand* and La Combattante II fast-attack craft *Joshan*, closed and engaged the American forces. Both were destroyed by a variety of PGMs fired at them.[21] Later, when the other Iranian frigate, the *Sabalan*, finally returned to the fight, it was disabled by an LGB.[22] Having exacted the desired toll on the Iranians, Secretary of Defense Frank C. Carlucci halted the operation before this vessel, too, was sunk.[23]

Operation SOUTHERN WATCH (13, 17–18 January 1993)

Operation SOUTHERN WATCH, which commenced in August 1992, created a "no-fly zone" below the 32nd parallel in Southern Iraq.[24] Designed to protect

16 Palmer, *On Course*, 116–33.

17 J.B. Perkins III, "Operation Praying Mantis: The Surface View," *Proceedings* 115.5 (May 1989): 68.

18 Palmer, *On Course*, 133, 142.

19 Ibid., 142.

20 For a short summary of the surface force action, see Perkins III, "The Surface View."

21 Palmer, *On Course*, 143–4; Huchthausen, *Splendid Little Wars*, 108; Perkins, "The Surface View," 69.

22 For a short summary of the air action, see Bud [Arthur N.] Langston and Don Bringle, "Operation Praying Mantis: The Air View," *Proceedings* 115.5 (May 1989): 54–60.

23 Palmer, *On Course*, 145.

24 In Northern Iraq another no-fly zone was patrolled under the auspices of Operation PROVIDE COMFORT.

the Iraqi Shia in the southern part of the country from attack, the operation was relatively quiet for the first few months following its inception. This all changed when Saddam Hussein began testing the coalition at the end of 1992, just prior to President Bill Clinton taking office. On 27 December, a USAF F-16 shot down an Iraqi MiG-25 "Foxbat" that had deliberately entered the prohibited airspace.[25] The Iraqi's continued to test the coalition forces into early January 1993 with more flight incursions into the no-fly zone, some cross-border ground incursions near Umm Qasr, and by the introduction of new surface-to-air missiles (SAMs) into the no-fly zone.[26] While little in the way of diplomatic efforts was going on out of the public eye on this issue, public rhetoric from both sides was pushing the situation towards a potential armed confrontation.

While tensions were increasing with regards to the no-fly zone, the Iraqis were simultaneously causing problems with the UN-mandated weapons inspection program. As tensions continued to escalate, planners in Riyadh and elsewhere were developing a series of response options for the president. Three punitive strikes were eventually launched into Iraq. Two of these were fixed-wing air strikes conducted in response to the no-fly zone violations, while the third was a TLAM strike near Baghdad ordered in response to the difficulties the UN was experiencing with carrying out their inspection regimen inside Iraq.

On 13 January U.S., British, and French aircraft attacked a series of six IADs-related target areas with somewhat mixed results. As White records, "According to Pentagon officials, the aircraft struck only 16 of 32 specific target points because cloud cover in the region limited target acquisition and employment of laser-guided munitions."[27]

Four days later a second Iraqi fighter was shot down. Later that night, the Navy launched its first post-DESERT STORM volley of TLAM in response to Iraq's continued refusal to comply with the UN inspection mandate. Forty-five missiles were launched at the Zaafaraniyah nuclear fabrication facility, which was suspected of being engaged in nuclear weapons development. While one missile impacted the Al-Rashid Hotel, killing three civilians, the strike was very successful as the facility was severely damaged.[28]

---

25 Paul K. White, *The Washington Institute for Near East Policy Military Research Papers*, vol. 2, *Crises After the Storm: An Appraisal of U.S. Air Operations in Iraq Since the Persian Gulf War* (Washington, DC: The Washington Institute for Near East Policy, 1999), 19–21.

26 Reuben E. Brigety II, *Ethics, Technology and the American Way of War: Cruise Missiles and US Security Policy*, Contemporary Security Studies (New York: Routledge, 2007), 76–8.

27 White, *Crises After the Storm*, 23; Brigety II, *Ethics, Technology and the American Way of War*, 79.

28 *RGM/UGM-109B/C Tomahawk*, Jane's Naval Weapons Systems, Issue 26 (April 1998), 2; White, *Crises After the Storm*, 24; Brigety II, *Ethics, Technology and the American Way of War*, 80.

The following day, 75 coalition aircraft returned to targets attacked the week earlier. This time, the strike was considered a great success. Using a combination of PGMs and unguided munitions, the strike rendered "Iraq's air defense in the South virtually inoperable."[29] Following these attacks, and with an explicit acknowledgment to the imminent change in American Administrations, Saddam relented. By the end of January the UN inspections had resumed and there were, for some time to come, no further violations of the no-fly zone.

Operation BUSHWACKER (26 June 1993)

The first use of precision-guided weapons, in this case TLAM, by the Clinton Administration occurred just six months later. In April, coincidental to a planned trip to Kuwait by former president George Bush, a group of Iraqis smuggled a cleverly-designed car bomb, liquor, and weapons across the border. Fortunately, the group was caught before its plan could be put into action and one of the perpetrators confessed during interrogation.[30] The discovery of the plot to assassinate the former president put into motion a highly classified planning operation.

Eventually, President Clinton himself made the final target selection, designed to punish the Iraqis for what was called by the then-U.S. Ambassador to the United Nation, Madeline Albright, "a direct attack on the United States."[31] Two U.S. Navy ships, one in the Red Sea and one in the Persian Gulf, targeted 23 missiles against the Iraqi Intelligence Service (*Mukhabarat*) Headquarters building. The strike was successful and the headquarters building was destroyed.[32] Because the strike was conducted on the Western weekend, Ambassador Albright actually had to visit Iraq's UN representative at his personal residence to inform him that the strike was underway.[33] Benjamin Lambeth believes that the origins of "cruise missile diplomacy" can be traced to this first non-combat firing of TLAM into Iraq.[34]

After the four air strikes in January and June 1993, the next significant confrontation with Saddam did not occur until October 1994. At that time, in response to the movement of thousands of Iraqi troops towards the border with Kuwait, the United States launched Operation VIGILANT WARRIOR. In the first eight days of this action, an aircraft carrier, some 350 aircraft and 28,000 combat troops were deployed. In this crisis, rapid action on the part of the U.S. appeared to stave off any need for offensive military action. Saddam returned his troops to garrison.[35] While the next two years were to be relatively quiet, this began to change in the summer of 1996.

---

29 White, *Crises After the Storm*, 24.

30 Brigety II, *Ethics, Technology and the American Way of War*, 88–9.

31 Brigety II, *Ethics, Technology and the American Way of War*, 93.

32 *RGM/UGM-109B/C Tomahawk*, 2; White, *Crises After the Storm*, 26.

33 Albright and Woodward, *Madam Secretary*, 273.

34 Lambeth, *NATO's Air War*, 179.

35 White, *Crises After the Storm*, 28–33.

Operation DESERT STRIKE (3–4 September 1996)

Although dealing with significant internal problems at home, on 31 August 1996 Saddam Hussein ordered Iraqi forces to attack the Northern Iraq city of Irbil with between 30,000 and 40,000 troops and associated armor and artillery. Two rival Kurdish factions, Massoud Barzani's Kurdish Democratic Party (KDP) and Jalal Talibani's Patriotic Union of Kurdistan (PUK), had fought for decades over control of Iraqi Kurdistan. Barzani, whose military fortunes had waned during the preceding several months, had reportedly asked for Saddam's help to avoid a decisive defeat at the hands of the PUK. Saddam obliged, and in less than 48 hours, Irbil was under Iraqi/KDP control.[36]

While the coalition had retaliatory air strikes planned and ready to execute, in the first taste of what some have subsequently called "sanctions fatigue," "Turkey, Jordan, and Saudi Arabia denied the U.S. request for launching coalition air strikes against Iraq from their territory."[37] This action left the Americans with few viable options with which to try to coerce Saddam. One option that was available, relying on precision-guidance *and* not requiring host nation permission, was a cruise missile attack.

As executed, Operation DESERT STRIKE was a two-day, two-wave assault against Iraq. Paradoxically, while trying to dissuade actions in Northern Iraq, the two cruise missile strikes were actually launched against air defense sites in Southern Iraq.[38] Secretary of Defense William Perry explained that these actions were designed to protect coalition pilots in what was now a newly expanded no-fly zone.[39]

The 3 September strike consisted of 13 CALCM, launched from two Guam-based B-52s, and 14 TLAM, launched from two U.S. Navy surface combatants, targeted against 15 individual targets. Following an initial battle-damage assessment (BDA), a follow-on strike consisting of 17 TLAM, launched from three surface combatants and one submarine, was targeted against four of the original targets.[40] Following the action, General Joseph Ralston, then-Vice Chairman of the Joint Chiefs of Staff, characterized the strikes as a success, but was reluctant to provide specific details.[41] The Iraqis quickly rebuilt the sites.

---

36 White, *Crises After the Storm*, 38–40; Brigety II, *Ethics, Technology and the American Way of War*, 115–17.

37 White, *Crises After the Storm*, 40; Brigety II, *Ethics, Technology and the American Way of War*, 118.

38 There were few other options. See Brigety II, *Ethics, Technology and the American Way of War*, 118–19.

39 William J. Perry, Secretary of Defense, Press Conference (Washington, DC, 3 September 1996); White, *Crises After the Storm*, 41; "Signals: US Navy Launches Tomahawk Strikes at Iraq," *Jane's Navy International* 101 no. 8 (1 October 1996): 8.

40 White, *Crises After the Storm*, 40–41; "Tomahawk Strikes," 8.

41 Joseph Ralston, Vice Chairman, JCS, Press Conference (Washington, DC, 4 September 1996); Later reporting indicated that most of the targets were damaged. Kenneth

Operation INFINITE REACH (20 August 1998)

Operation INFINITE REACH marked the first time that TLAMs were targeted into two countries simultaneously and, perhaps more significantly, marked the first time that U.S. authorities targeted a non-state actor with the weapon. The two-pronged attack struck six suspected terrorist training camps belonging to Osama bin Laden in Afghanistan and the *Al-Shifa* (Health) Pharmaceutical Factory in Sudan. As Thomas Mahnken notes, the operation was both retaliatory and preemptive.[42] The operation into Afghanistan was launched as punishment for the twin bombings of the U.S. embassies in Kenya and Tanzania on 7 August 1998 and also to "send a forceful message on the global reach of the USA to international terrorist organizations."[43] The operation into Sudan was designed to destroy the factory, which had been suspected by the Americans of producing precursor chemicals for the nerve agent VX.[44]

As President Clinton told the nation in after the bombing,

> I ordered this action for four reasons: First, because we have convincing evidence these groups played the key role in the embassy bombings in Kenya and Tanzania. Second, because these groups have executed terrorist attacks against Americans in the past. Third, because we have compelling information that they were planning additional terrorist attacks against our citizens and others with the inevitable collateral casualties we saw so tragically in Africa. And, fourth, because they are seeking to acquire chemical weapons and other dangerous weapons.[45]

Some, however, questioned the strike's timing as it occurred just three days after President Clinton admitted to his relationship with White House intern Monica Lewinsky. Similar questions would arise again four months later when the president ordered another set of strikes into Iraq.[46]

Controversy also quickly engulfed the decision to bomb the pharmaceutical plant, which killed the night watchman and his family.[47] Many reported that the

---

H. Bacon, Assistant Secretary of Defense (Public Affairs), Press Conference (Washington, DC, 6 September 1996).

42 Thomas G. Mahnken, *Technology and the American Way of War Since 1945* (New York: Columbia University Press, 2008), 188.

43 Bryan Bender, "US Double Strike: 'A New Ball Game'," *Jane's Defence Weekly* 30.8 (26 August 1998): 8.

44 Bender, "US Double Strike," 8.

45 William Jefferson Clinton, President, Statement by the President (Martha's Vineyard, 20 August 1998).

46 Madeleine Albright reacts harshly to that criticism. Albright and Woodward, *Madam Secretary*, 368.

47 Werner Daum, "Universalism and the West: An Agenda for Understanding," *Harvard International Review* (Cambridge, MA) 23.2 (Summer 2001): 19. Other sources

plant had nothing to do with nerve agents although the Clinton Administration held firm on the accuracy of its pre-strike assessment.[48] While officials were initially very reluctant to discuss the details of the operation, information has since slowly emerged. Sixty-two of the TLAM were launched against the six suspected camps in Afghanistan.[49] Another six or so were reportedly launched against the plant.[50] Reaction around the world was mixed and, in the end, it seems little was accomplished beyond "sending a message." One author categorizes them as "the costly, yet apparently ineffectual, TLAM strikes launched ... by the administration against presumed assets of the terrorist Osama bin Laden in Sudan and Afghanistan."[51]

Operation DESERT FOX (16–19 December 1998)

Following Operation DESERT STORM, the United Nations Special Commission (UNSCOM) was formed by UNSCR 687 and tasked to "carry out immediate on-site inspections of Iraq's biological, chemical and missile capabilities, based on Iraq's declarations and the designation of any additional locations by the special commission itself."[52] From the start, discord existed between the Iraqi government and the UN inspection teams, resulting in one of the earlier TLAM strikes noted. Between November 1997 and December 1998, four distinct crises between the two parties emerged. The first three were eventually deescalated; the final became Operation DESERT FOX.[53]

On 15 December, Richard Butler, then UNSCOM's chief inspector, issued a report to the United Nations Security Council which reported that "Iraq had not provided the full cooperation it promised ... and had in fact introduced new restrictions into the inspection process, ensuring no progress in the areas

report ten injured. Phinney, *Airpower Versus Terrorism*, 37.

48 See, James Risen, "To Bomb Sudan Plant, or Not: A Year Later, Debates Rankle," *New York Times* (New York), 27 October 1999, Late Edition—Final, A: 1; Phinney, *Airpower Versus Terrorism*, 35–6.

49 Jamie McIntyre, "'Moderate to Severe' Damage Seen at Suspected Bin Laden Camps," *CNN.Com* 13 January 1999 (accessed 29 December 2008): http://www.cnn.com/WORLD/asiapcf/9901/13/afghan.damage.photos.

50 Phinney, *Airpower Versus Terrorism*, 36. Without official government confirmation it is difficult to determine exactly how many missiles were used. Estimates range as low as 68 to as high as 79. Madeleine Albright, for one, reports the latter number. Albright and Woodward, *Madam Secretary*, 368.

51 Lambeth, *NATO's Air War*, 180.

52 UNSCR 687 (3 April 1991), para. 9.(b).(i). Available at: http://daccessdds.un.org/doc/RESOLUTION/ GEN/ NRO/596/23/IMG/NRO59623.pdf?OpenElement (accessed 23 December 2008).

53 White, *Crises After the Storm*, 49–58; Albright and Woodward, *Madam Secretary*, 277–84.

of disarmament and failing to account for its prohibited weapons programs."[54] While the timing was awkward, the vote to impeach him was pending in the House of Representatives, President Clinton ordered a "'strong, sustained series of air strikes' against Iraq that lasted four consecutive nights."[55]

The stated objectives of this short campaign were: to reduce Iraq's capability to produce WMD; to degrade strategic and tactical command-and-control facilities; to damage the industrial infrastructure used for smuggling gas and oil, and; to reduce Iraq's overall capability to threaten its neighbors in the region.[56] Over the next four nights around 100 targets were struck. In addition to the types of targets associated with the four objectives stated above, Republican Guard forces and air defense sites were also attacked.

In four days of bombing raids, primarily carried out by U.S. and UK forces, approximately 650 sorties were flown without any coalition losses. Additionally, 415 cruise missiles (325 USN/90 USAF) were launched. Congressional testimony by General Anthony Zinni, CINCENT at the time, later revealed that the strikes "hit and damaged more than 80 percent of the designated targets ... which Pentagon analysts assessed had set back Iraq's ballistic missile program one to two years."[57]

Madeleine Albright was similarly pleased, believing DESERT FOX "marked a turning point in what had become an international soap opera with the United States, the Security Council, and Saddam Hussein in lead roles."[58] Others, however, remained less than impressed. Nation notes that, "punitive air attacks were used to demonstrate political will but without seeking any kind of decisive strategic result."[59] Lambeth observes that, "It [the era of 'cruise missile diplomacy'] culminated in the three-day Operation Desert Fox, a mini-air operation that was executed against Iraq, to no significant consequence, at the very height of President Clinton's impeachment trial in December 1998."[60]

An Analysis of the Performance of Precision-Guided Munitions

As was the case with the preceding three case studies, the first level of analysis will examine the effectiveness and manner of use of precision-guided munitions employed in these strikes. These weapons were used in multiple theaters of operation and in different operational environments, both on land and at sea. One factor not present in the earlier case studies, but prominent here, is the compressed

---

54 White, *Crises After the Storm*, 58–9.

55 Ibid, 59.

56 Ibid.

57 Ibid., 60.

58 Albright and Woodward, *Madam Secretary*, 286.

59 Nation, *War in the Balkans*, 248.

60 Lambeth, *NATO's Air War*, 180.

planning time available. This factor, coupled with the continuing trend of ever greater public and governmental sensitivity to collateral damage and civilian casualties could have resulted in significantly negative results based on this study's methodology. However, as events transpired, that was not the case.

Part I: The Weapons and Their Employment

Although the seven contingency strikes examined spanned a period of more than a dozen years, the operational outcomes of these strikes are remarkably consistent. This is particularly significant because the circumstances in which the strikes took place, both in terms of the operational environment and in terms of the threat to the delivery platforms, varied considerably. During this period, there were also substantial improvements in precision-guided technology and an ever greater reliance on cruise missiles also emerged. While detailed post-strike analysis is not available for each of the sub-cases, enough information is available to determine the trend lines. In the final analysis, while the strikes themselves may have, at times, been seen as controversial, they all appear to have been remarkably discriminate.

Operational effectiveness Across the spectrum of these contingency strikes, precision-guided weapons continued to be highly accurate although certainly their performance was not perfect. In almost every case, however, the problems are not due to a fault with the weapons themselves, but with the delivery crews; the technology has from the very beginning proven very reliable. This was most apparent in the first sub-case, Operation EL DORADO CANYON.

Having to fly roughly 2,800 miles from their home base in England, before they could even attempt to release their weapons, presented significant logistical and operational challenges to the U.S. Air Force F-111 bombers. One of the most critical was the very real possibility that the strike force would be depleted by mechanical failure before it arrived on target. To a degree, that is just what occurred. Of the 12 USAF F-111s assigned to drop LGBs, five aborted. Four of these aborts were due to aircraft or weapon systems malfunctions; one did not drop, in keeping with the established ROE, because the crew couldn't properly identify their aimpoint.

The Libyan air defenses were also the most formidable of the seven strikes examined and one aircraft was hit by a missile, it is believed, and crashed. Despite these obstacles, five aircraft delivered their weapons close enough to the desired aimpoints to cause significant damage, although admittedly there were no direct hits. One aircraft missed its assigned target due to a navigational/target identification error, which resulted in civilian casualties and collateral damage.[61]

In Operation PRAYING MANTIS, on the other hand, with the exception of one weapon all PGMs delivered precisely impacted their designated targets.

61 Stanik, *El Dorado Canyon*, 184–91.

One Harpoon anti-ship missile over flew the wreckage of the Iranian ship *Joshan* because the vessel was already sinking when the weapon arrived.[62] Tomahawk cruise missiles also continued to demonstrate the accuracy first seen during Operation DESERT STORM.

In January 1993, 40 of 45 arrived at the Zaafaraniyah nuclear fabrication facility.[63] Three others landed inside the target complex but did no damage.[64] In the June attack launched in response to the plot to assassinate former president Bush, 19 of the 23 missiles launched struck the Iraqi Intelligence Service Headquarters building.[65] Three years later, TLAM again proved effective during Operation DESERT STRIKE.

While initially tight-lipped about specific results right after that strike, the Department of Defense later reported that six of eight SAM sites were hit, the other two had been moved prior to the strike. Of the seven IADs targets, six had some degree of damage, although the spokesman would not address specific aimpoints or missile effectiveness.[66] Other sources, however, reported that 29 of 31 arrived on target.[67]

While no publicly released information is available on the Sudan strike, press reports indicate it was executed with incredible precision. One account notes that, "The factory's walls fell inward while, not 10 feet outside them, 55-gallon drums remained standing. The roof came down, the assembly line was blown apart, and the candy factory that shares a wall with the plant was left intact."[68] Photos released in January 1999 showed significant damage to all three terrorist sites, although specifics regarding weapon performance have not been forthcoming.[69] Finally, with regards to Operation DESERT FOX, "According to Gen. Anthony Zinni, the 425 cruise missiles fired during Desert Fox 'far exceeded' the 85 percent accuracy standard."[70] Similar comments, recorded above, were also made by General Zinni with regards to the overall performance of aircraft delivered PGMs.

It is clear then, that against the two metrics of CEP and weapons functional effectiveness, the trend line continues to be positive for PGMs. Incremental improvements made to enhance the capability of American forces to deliver PGMs in adverse weather, primarily through the use of GPS, have not come

62 Palmer, *On Course*, 144; Perkins III, "The Surface View," 69; Langston and Bringle, "The Air View," 58–9.

63 *RGM/UGM-109B/C Tomahawk*, 2.

64 Brigety II, *Ethics, Technology and the American Way of War*, 79.

65 *RGM/UGM-109B/C Tomahawk*, 2; Brigety II, *Ethics, Technology and the American Way of War*, 92.

66 Bacon, Press Conference.

67 *RGM/UGM-109B/C Tomahawk*, 2.

68 Quoted in, Phinney, *Airpower Versus Terrorism*, 37.

69 McIntyre, "'Moderate to Severe' Damage Seen at Suspected Bin Laden Camps."

70 White, *Crises After the Storm*, 81; Duncan Lennox, "'Fox': The Results," *Jane's Defence Weekly* 31.2 (13 January 1999): 63.

at a significant cost in accuracy. Such an assessment does not imply perfection however. As demonstrated by one aircraft in the Libya strike, and by an errant missile in Iraq, technology and those who employ it are not universally infallible.

Manner of employment White notes that "The worldwide impact of almost instantaneous news reports and on-the-spot live coverage of events have increasingly made concerns of collateral damage and civilian casualties a high priority in any air campaign."[71] Despite this observation, given the hurried nature of some these contingency strikes it might be reasonably expected that for these types of missions less care would have been taken to ensure that PGMs were employed in a manner specifically designed to reduce human suffering. Fortunately, the evidence does not support such a negative conclusion.

In the first of these contingency strikes, Operation EL DORADO CANYON, "To inflict less damage on noncombatants, American planners chose targets that had fewer homes nearby—in case the bombs missed their target." Vice Admiral Frank Kelso, USN, the officer in overall charge of the operation, limited his pilots to a single pass and mandated that in the event of anything less than 100-percent target identification and a fully functional precision-guidance system, pilots were prohibited from releasing their ordnance.[72] Interestingly, given the post-DESERT STORM bias towards using them in such situations, an early decision was made not to consider the use of TLAM due to its "strategic roles," and over concern that any unexploded missiles would end up in the hands of the Soviets.[73]

By attacking the intelligence headquarters building at night, the planners of Operation BUSHWACKER ensured that, while the political statement was made, the lives of innocent government employees, who would be working there during the day, were spared. Unfortunately, during this strike three TLAM missed the target area and impact nearby homes, killing eight civilians and wounded some 12 others.[74] During Operation PRAYING MANTIS, although not directly involving the portion of the strike which used PGMs, enemy personnel onboard the two GOSP platforms were warned of the impending attack, then given time to leave before firing commenced. Interestingly, there were also times when the planners did not wish to limit potential casualties

During that portion of Operation INFINITE REACH launched against the six suspected terrorist training camps in Afghanistan, the U.S. "designed the attacks to maximize casualties, timing the strike to coincide with a scheduled gathering of suspected terrorists and their leaders. ... U.S. officials believed about 100

71 White, *Crises After the Storm*, 80.

72 Phinney, *Airpower Versus Terrorism*, 18; Stanik, *El Dorado Canyon*, 150–52.

73 Phinney, *Airpower Versus Terrorism*, 16; Stanik, *El Dorado Canyon*, 111.

74 Brigety II, *Ethics, Technology and the American Way of War*, 92.

terrorists-in-training were killed."[75] The difference here, of course, is that the planners viewed these individuals as enemy combatants, not innocent civilians.

But for those strikes where explicit attention was paid to minimizing human suffering, the vast majority, the trend line is overwhelmingly positive. In Libya, the very few errant weapons, as earlier recorded, killed 37 civilians, and injured another 130.[76] In Iraq in January 1993, just 19 Iraqis were reported killed on 13 January by the first of the LGB strikes, while only around 30 were killed by the second strikes on 18 January.[77] As noted, eight civilians were killed by Operation BUSHWACKER. Even the much larger Operation DESERT FOX resulted in few civilian casualties. The Iraqis themselves reported only 42 civilians dead and 96 wounded.[78] While any innocent deaths are regrettable, these statistics show that the level of human suffering is far lower than in past conflicts, where upwards of tens of thousands of civilians were killed in a single night's attack.[79] Clearly this is a positive trend.

Part II: The Question of Unintended Consequences

This portion of the analysis normally relies upon three separate lines of inquiry when exploring concerns over second order effects. However, when examining these contingency strikes, only one line of inquiry emerged as offering any useful information for the study. While it is possible that a contingency strike could result in the rapid capitulation of an enemy, perhaps in the form of a popular uprising resulting in the overthrow of the government, this was not the objective, nor did it occur, in any of the contingencies in this study. Similarly, while a contingency strike might act as a catalyst for an asymmetric response, one scenario might be that of a retaliatory terrorist attack on American soil following a TLAM strike, this too did not occur. Unfortunately, as was seen in DESERT STORM, long-term suffering can result even when a given strike was particularly discriminate.

Long term suffering Such was the case with the Sudanese portion of Operation INFINITE REACH. Avoiding the debate of whether or not the pharmaceutical plant played any actual role in developing precursors for weapons of mass destruction, it did produce lifesaving medicine for the Sudanese people. Its destruction did, in fact, cause human suffering that otherwise could have been avoided.

75 McIntyre, "'Moderate to Severe' Damage Seen at Suspected Bin Laden Camps."

76 Werrell, *Chasing the Silver Bullet*, 154.

77 White, *Crises After the Storm*, 23–4.

78 Ibid., 60.

79 For a quick comparison of deaths resulting from a number of selected air campaigns in the twentieth-century see Ward Thomas, *The Ethics of Destruction: Norms and Force in International Relations* (Ithaca, NY: Cornell University Press, 2001), 169.

Werner Daum, who at the time of the strike was Germany's ambassador to Sudan, records that the *Al-Shifa* factory, destroyed by the INFINITE REACH strike, was responsible for producing 20 to 60 percent of the pharmaceuticals in Sudan and, perhaps more critically, 100 percent of the country's intravenous liquids. He noted that it took "more than three months for these products to be replaced with imports."[80] In his opinion, while acknowledging that it is hard to assess, "tens of thousands seems a reasonable guess" to the question of how many Sudanese died due to the destruction of the factory.[81] Thus, while the trend line across all seven sub-cases is positive, this must necessarily be tempered by the Sudan experience.

Part III: The Question of Teleology

Similarly the final portion of the analysis, which is focused on "ends," normally rests on two lines of inquiry. It is clear from the literature that, for these seven examples, using ground forces was never a serious consideration. In at least two cases this was quite explicit. For Operation EL DORADO CANYON, as Phinney records, "Air power was the only option seriously considered."[82] Similarly, during the planning leading up to the INFINITE REACH strike, a similar mindset was in place. At a 12 August meeting, less than a week after the twin embassy bombings, "General Shelton told the president that he had considered attacks by ground forces and conventional aircraft but had ruled both out due to the risk they placed upon the American combatants."[83]

However, the fact that ground troops were not employed made no difference in terms of human suffering in these instances. The very nature of these no-notice contingency operations is that they are very closely-held operations. While an enemy may surmise that a U.S. armed response is possible, given the state of tension with the United States, the element of surprise, particularly at the tactical level, is significant. As a result, there simply is no evidence that would support a trend determination in either a positive or negative direction. The second line of inquiry is, however, worthy of exploration.

Greater willingness to resort to lethal force This last line of inquiry seeks to determine whether having the ability to strike with relative impunity, using precision-guided weapons, actually encourages leaders to resort to force more often, before traditional *jus ad bellum* criteria have been satisfied. The fact that for each of these contingency operations the resort to military force was not discussed openly in either a UN or NATO forum, as was done before Operations DESERT STORM, DELIBERATE FORCE, and ALLIED FORCE, but planned in

80 Daum, "Universalism and the West," 19.

81 Ibid.

82 Phinney, *Airpower Versus Terrorism*, 15.

83 Ibid., 34.

secrecy would seem, at least initially, to indicate a negative trend according to the framework. But, as is often the case, the picture is more complicated.

The most common criticism is that these operations were "illegal," lacking the elements of legitimate authority, from the just war tradition, or legal justification. Were this true, then the initial trend determination would have to be accepted. However, while these operations lacked the open and robust debate seen at other times, the senior leadership of the U.S. did not think they were without a legal basis to conduct the operations and these bases were discussed before the operations.

As early as January 1986, the State Department's legal counsel had determined that "according to Article 51 of the United Nations Charter terrorism constituted 'armed aggression' and that military force could be used in self-defense."[84] This underpinned the decision to use force in Libya. The Article 51 argument was also used to justify the Operation BUSHWACKER TLAM strike.

On 27 June, Ambassador Albright remarked, "this [assassination attempt] was a direct attack on the United States, an attack that required a direct United States response. ... We responded directly, as we are entitled to, under Article 51 of the United Nations Charter, which provides for the exercise of self-defense in such cases."[85] Similarly, President Clinton's comments regarding the INFINITE REACH strikes, recorded above, reflect an Article 51-like rationale. As Finn notes, "Whilst both operations [Libya and Afghanistan/Sudan] were subject to international criticism neither were subject to legal challenge nor to Security Council condemnation."[86]

With regards to the series of actions in Iraq, it has been persuasively argued that those operations, while not authorized by new UNSCRs, took place under earlier resolutions that were still in effect. UNSCR 678 had previously called for the restoration of "international peace and security" in the region and authorized the use of force. That mandate had never been lifted and, in fact, had been reaffirmed in UNSCR 949 (15 October 1994). It must be acknowledged, however, that critics could easily criticize the DESERT STRIKE operation, as the goal of punishing Iraqi actions in Northern Iraq by temporarily suppressing air defenses in the South seems incongruent.

Finally, the framework admitted of the possibility that a trend line, first seen as negative, could be seen in a different light if the action taken alleviated later human suffering. While difficult to quantify the net reduction of future suffering, as was earlier noted, after the strike in Libya terrorist attacks dropped off markedly for a number of years although the downing of Pan Am Flight 103 over Lockerbie, Scotland cannot be ignored. Similarly, following action in Iraq in 1993, the no-fly

---

84 Stanik, *El Dorado Canyon*, 109.

85 Brigety II, *Ethics, Technology and the American Way of War*, 93.

86 Chris Finn, "The Broader Implications of the Increasing Use of Precision Weapons," *Air Power Review* 4.1 (Spring 2001): 43.

zone became much quieter and the UN inspectors were allowed to continue their inspection regimen.

This study has now examined 11 cases where PGMs played some role. At first, PGMs were a small part of much larger operations. As technologies improved, and greater emphasis was placed on keeping civilian casualties and collateral damage to a bare minimum, they came to play a bigger, perhaps even dominant role in military operations. The study will next turn to a summary of the findings, including some reflections on contemporary operations, and the conclusions that have been drawn from this effort.

Chapter 8
In The Final Analysis: Summary with Contemporary Reflections and Conclusions

> In the air, we must be able to strike from across the world with pinpoint accuracy—with long-range aircraft and perhaps with unmanned systems
>
> George W. Bush

This study began with the observation that strategic culture, ways of warfare, and technological innovations are social constructs. Different societies will, given the same technologies and circumstances, seek different solutions based on heritage, values, and prior experiences. For the United States, one manifestation of this process is the increasing reliance on the use of precision-guided munitions to wage war. By doing so, the U.S. seeks to use such weapons to compel the adversary to bend to its will using ever-*decreasing* levels of violence and attendant human suffering.

Indeed, this is not the first time a new weapon has been seen to hold such promise; gunpowder and artillery are but two examples from earlier times. Two of the measures of effectiveness often employed in contemporary times are those of civilian casualties and collateral damage, the latter a euphemism for the destruction of buildings, infrastructure etc. that were not the intended military target. But has this positive potential of PGMs been realized, where earlier weapons technology fell short? Here, then, lies the study's core question: Do PGMs mitigate suffering in war and have these weapons changed the way decisions regarding war and peace have been made? To seek answers to this seemingly simple, but multi-faceted question, a comprehensive evaluative framework was developed, which adopted a multi-level, longitudinal approach.[1] As its foundation, the framework rests upon the Just War Tradition, including emerging *jus post bellum* considerations. However, while one hopes that any valid ethical tradition is complete, comprehensive, non-contradictory, and livable, to ensure inclusiveness, completeness, and to gain greater insight into the roots of the tradition, a number of other ethical, legal, and philosophical viewpoints were engaged.

The study used four case studies to examine eleven occasions where force was used by the American government to further national interests, sometimes alone and at other times in partnership with other nations. These cases met two key criteria; they matched the study's definition of war and they relied upon, at least in

1 The framework was designed such that other technological innovations could be measured using the instrument by simply altering the metrics used. With minor modification the lines of inquiry are enduring, the metrics situationally dependent.

part, the use of PGMs. This chapter summarizes the findings across these diverse cases while injecting commentary on more contemporary, but less historically settled operations, and outlines the logical conclusions drawn from those results. It closes with some observations on the future of PGMs and warfare.

Summarizing the Findings

The cases selected included all major, and most of the lesser, combat operations the United States engaged in between 1965 and 1999. These ranged from the war in Vietnam, where LGBs were used in the later stages of a much wider conflict, to the overwhelming coalition victory in Operation DESERT STORM, to the twin NATO operations in the Balkans, and to Operation INFINITE REACH, where TLAM was used alone in an operation. Despite the differences in time, precision technology available, actual weapons utilized, and operational environment, the results at each level of the analysis remained remarkably consistent across the longitudinal axis of the study. As before, each level will be examined separately.

Part I: The Weapons and Their Employment

Since their first use in combat in Southeast Asia, the incredible potential of modern PGMs captured the interest of both the military professional and the civilian leadership of the United States. While Clodfelter records, "For 1971 smart bombs reportedly accounted for less than one percent of the total weight of munitions dropped in Indochina,"[2] PGM usage expanded rapidly in conflicts thereafter. Twenty years later, in Operation DESERT STORM, PGMs accounted for about 8 percent of those munitions employed.[3] In later operations, including Operation DELIBERATE FORCE in Bosnia, PGMs accounted for more than 90 percent of the ordnance used by American forces. This greater trust in the accuracy and reliability of PGMs appears to have been justified by the evidence available.

Operational effectiveness From the earliest developmental tests observed by Colonel Joseph Davis, in the mid-1960s, LGBs have consistently demonstrated accuracies, as measured by their CEP, orders of magnitude lower than those previously achieved by conventional unguided weapons. An important caveat here is that this was, and is, true only when the aircrews were properly trained and no weather conditions, such as clouds, or smoke, interrupted the laser-designation during the weapons flight. With LGB CEPs in Vietnam hovering around 30 feet, and more than 50 percent of the weapons being observed as direct hits, the initial impressions made were exceptionally positive. As Major General Eugene L. Hudson, then-7th Air Force Director of Intelligence, observed in Vietnam, "Laser-

2 Littauer and Uphoff, eds., *Air War in Indochina*, 156n.23.

3 Keaney and Cohen, *GWAPS*, 226.

guided bombs ... revolutionized tactical bombing."[4] Electro-optical weapons also fared very well during this conflict.

After the war, improvements were continually made to LGB/EO technologies including improved guidance and tactics such as expanding the ability of aircraft to deliver LGBs from low altitude by using a loft maneuver. In the 1980s and early 1990s, LGBs, along with other PGMs, performed well in both Libya and against Iranian naval vessels in the Persian Gulf region. Martin Van Creveld observes, "As each new generation of weapons entered service, the lethality of combat increased by leaps and bounds. The 1991 Gulf War seemed to justify the investment insofar as ... some 90 percent of all 'smart' bombs launched were said to have hit their targets."[5]

Such a capability led Colonel John Warden, architect of the early INSTANT THUNDER planning, to declare:

> The second huge revolution is the military technological revolution which is a combination of tremendous computational power, increasingly unlimited bandwidth and a number of other technologies—the key one of which is precision. Those things came together for the first time in the Gulf War where we were able to see what they could actually do as we watched F-117s flying over Baghdad dropping single bombs and hitting exactly what they were supposed to. This in my mind is not *a* revolution in military affairs but is *the* first genuine military technological revolution ever because for the first time we now have a *conceptually* different way to wage war.[6]

But as earlier noted, this was a bit of an illusion, as far fewer PGMs were actually used during the Gulf War than most of the world believed. However, their positive, casualty-reducing performance, particularly in highly populated areas, was indeed impressive.

The first operational use of TLAM and CALCM cruise missiles also occurred during Operation DESERT STORM. These early variants did very well in their debut, demonstrating CEPs similar to that of LGBs (15–30 feet) and high degrees of weapon's functional effectiveness, reported as greater than 80 percent for both.[7] Warden also noted that "In the past, war was defined by misses. This war with precision was defined by hits."[8] However, while the success of LGBs was hailed, laser and electro-optical technology had their operational limitations.

As Sandy Erwin noted later, "laser-guided weapons often are the preferred choice in these operations, but lasers require a direct line of sight to the target, which

4 Quoted in Clodfelter, *Limits of Airpower*, 159.

5 Van Creveld, *Technology and War*, 269.

6 Warden, *Planning to Win*, 8.

7 Schwamb et al., *Desert Storm Reconstruction Report, Vol. II*, xv, 8-3, provides the TLAM data.

8 Warden, *Planning to Win*, 8.

often can be obstructed by bad weather, smoke or buildings."[9] This problem was witnessed in both DESERT STORM, the Balkans, and in post-DESERT STORM strikes into Iraq. To help alleviate this problem, thereby expanding the range of options planners would have for the employment of PGMs, the military turned to Global Positioning System (GPS) navigation. This technology, utilizing an orbiting constellation of navigation satellites, does, perhaps counter-intuitively, result in a slightly higher CEP, although the 40-foot CEP advertised is still a vast improvement over earlier weapons. JDAM, fundamentally a bolt-on kit incorporating GPS receivers and other hardware that turns ordinary unguided general purpose bombs into PGMs, was hurried into battle in Kosovo and performed very well with over 90 percent of the weapons striking their assigned targets.[10]

But as Van Creveld correctly observes, "there is no weapon but that has its limitations and no technology so perfect that it cannot, in principle at any rate, be countered with the aid of appropriate organization, training, and doctrine."[11] This has been a concern with JDAM and other GPS-guided weapons due to the ready availability of inexpensive GPS jammers. U.S. Air Force General Mike Moseley, for example, was worried about the Russian-made GPS jammers that the Iraqis had mounted on towers to try to impede the Americans from using their JDAM satellite-guided bombs.[12] These towers were, interestingly enough, obliterated by F-117s. "In one of the war's ironies, the GPS jammers were put out of action by GPS bombs."[13]

TLAM, whether guided by the older TERCOM/DSMAC system or newer GPS technology, has continued to prove highly effective. Across different theaters and operational environments, the missile has functioned as designed, arriving on target, within 30 feet of its desired aimpoint, more than 80 percent of the time. As the framework has been formulated, evaluation of this first line of inquiry relies upon two metrics, CEP and weapons functional effectiveness. The results are that across all four case studies, with the few minor exceptions noted in the earlier narration, the trend lines observed have remained strongly positive. While firm data is difficult to obtain, press reports from Afghanistan (2001-present), Iraq (2003–2011), and even Libya (2011) indicate this trend continues. As one observer noted, "Fern-bar critics notwithstanding, American high-tech weapons perform splendidly."[14]

Manner of employment As prominent defense analyst Norman Friedman writes, "Since 1945, and particularly since the end of the Cold War, the U.S. view has been that direct attack against enemy civilians is an evil to be avoided. The implication is that the objective in war is simply to stop an enemy government from some

9 Erwin, "Air War Over Iraq," 68.

10 Gillespie, *Precision Guided Munitions*, 210.

11 Van Creveld, *Technology and War*, 230.

12 Gordon and Trainor, *COBRA II*, 322.

13 Ibid., 324. GPS weapons do not need continuous updates from the satellites to be accurate.

14 Reed, "Today's Weapons Overly High-Tech," C11.

objectionable act."[15] Thus, having concluded that PGMs are indeed as accurate and operationally effective as anticipated, a determination based on the study needs to be made as to whether these precise weapons are indeed being used in a manner designed to reduce human suffering. This is important because a positive determination demonstrates that American planners are actually aligning military effort to what their political leaders' desire, victory with a minimal amount of civilian casualties and collateral damage.

It must be remembered, from the early Vietnam case study, that American civilian and military leaders specifically addressed reducing unnecessary suffering even before the first LGB was dropped in combat. As Ward Thomas noted, in Vietnam, the "target selection process itself ... was designed so that almost as much weight was given to avoiding civilian casualties as to military effectiveness, and sometime more."[16] Furthermore, "It is clear from the restrictions on U.S. air crews, as well as from accounts of the political processes from which those restrictions emerged, that priority was placed not simply on the avoidance of *unlawful* civilian casualties but the avoidance of civilian casualties, period."[17]

Such restrictions, in a variety of forms such as "no-strike" lists, restrictive ROE, and mandatory run-in headings, were later put into place in operations over Libya, Bosnia, Kosovo, and, to a lesser degree, in Iraq. In fact, as evidenced by the complaints sometimes aired by pilots, the senior leadership seemed so concerned about the political ramifications of an "incident" that they appeared, at times, to be more concerned with civilian casualties than possible air crew losses. Such concerns aside, it is clear that the combination of the new, precise weapons and careful operational employment has had a significant, positive effect on reducing human suffering in war.

For example, Guenter Lewy, in *America in Vietnam*, concludes that despite the popular mythology that emerged from this period, "detailed examination of battlefield practices reveals that the loss of civilian life in Vietnam was less great than in World War II and Korea and that concern with minimizing the ravages of war was strong."[18] While his statement might be historically accurate, it must be acknowledged that until PGMs came into use, there was tremendous human suffering in Southeast Asia. However, during the LINEBACKER II operation where LGBs played a much larger role, civilian casualties dropped considerably, beginning a marked decline that continues to the present.

During this final eleven-day air operation, 1,318 civilians were killed in Hanoi, while another 305 were killed in Haiphong. This is far below the many thousands that were once routinely killed in bombing raids during World War II and Korea. During the longer DESERT STORM air campaign, roughly 2,278 civilians were killed. While this was a larger number of deaths overall, the campaign was more

15 Friedman, "Too Precise?" 4.

16 Thomas, *Ethics of Destruction*, 153.

17 Ibid., 155.

18 Lewy, *America in Vietnam*, vii.

than four times longer. General Buster Glosson and his planning teams worked hard to ensure that such casualties were kept low and that damage to the nation's infrastructure was not long-term. In one indication that this effort was successful, William Arkin later reported that, "the air war was 'clean on a strategic level,' and that he could find no evidence of indiscriminate attacks on cities or civilians ... or extensive collateral damage of civilian structures near targets."[19]

In the Balkans, similar planning concerns were expressed by those in senior leadership positions during the operations in both Bosnia and Kosovo. Again, this attention to reducing human suffering was effective. In Bosnia, Milošević later admitted that only 25 Serbs were killed during the air campaign.[20] As recorded earlier, during the much longer ALLIED FORCE operation in Kosovo, only about 500 civilians were killed by the allied bombing effort. In the seven contingency operations examined, known civilian casualties ranged from eight killed during Operation BUSHWACKER, to 42 killed during the longer DESERT FOX.[21] In June 2006, as but one more contemporary example, an F-16 precisely destroyed the safe house of al-Qaida leader Abu Musab al-Zarqawi, and its two inhabitants, without damaging surrounding structures.

It is also notable that, in addition to lower level of civilian casualties and associated collateral damage, the era of the PGM also witnessed a sharp drop in military deaths, on both sides of the battle line. This was, in fact, the first benefit realized by the use of LGBs in Vietnam. Greater precision meant fewer aircraft deep within the enemy's air defense system. Fewer aircraft inside that "threat envelope" resulted in fewer downed aircraft and fewer dead or captured airmen. As the trend towards the use of greater stand-off weapons has continued, so too has this pattern. Most of the operations studied had only one or two allied aircraft lost and, as noted in cases such as Bosnia, very few enemy casualties. More contemporarily, operations with Unmanned Aerial Systems (UAVs) effectively take the human out of the cockpit eliminating even the possibility of aircrew loss. Against the framework's metrics for this line of inquiry, these are all positive trends.

Part II: The Question of Unintended Consequences

A review of the evidence obtained at the first level of analysis indicates that PGMs have indeed had a positive influence by markedly reducing human suffering in war. So far, however, the analysis has focused on events that occurred during the period of actual combat operations, in other words, principally *in bello* considerations. The summary next turns to the three lines of inquiry that comprise the second level of analysis. Here *post bellum* considerations are taken into account to see if any evidence of second order effects, or unintended consequences, gleaned from the case studies would add support to, or fundamentally alter, the earlier preliminary

19 Arkin, "Tactical Bombing of Iraqi Forces," 62.

20 Conversino, "Executing Deliberate Force," 161.; Beale, *Bombs Over Bosnia*, 37.

21 White, *Crises After the Storm*, 60.

conclusion. Here the summary will also address some of Jacques Ellul's core concerns.

Rapid capitulation with unresolved tensions The intellectual process that underpinned the development of the evaluative framework for this work occurred before the selection of the actual case studies. This was done, by design, to keep the measurement instrument from becoming too heavily influenced or narrowly focused so that the framework could be used to evaluate other technologies. As a result, this line of inquiry was actually modified during the study to add greater value to the framework. The original intra-state focus of this line of inquiry had to be expanded to include an inter-state dimension based on the knowledge gained from Operation DESERT STORM.

In that case, while the war was quick, and victory rapidly declared, the Iraqis never actually capitulated. Instead, a ceasefire was unilaterally declared by the side that was winning at a time, a ceasefire that was to remain in place until the U.S.-led invasion in 2003. In fact, removal of Saddam Hussein was never a stated goal of the coalition in 1991. Thus, while résistance movements certainly emerged in both the north and the south, the resultant human suffering cannot logically be attributed to PGMs or the manner in which they were employed. However, within the expanded *inter*-state context of this portion of the framework, the destruction of significant portions of Saddam's army did reduce his ability to threaten his enemies, which likely resulted in few instances of aggression for years to come, which can be seen as a positive trend.

In both Bosnia and Kosovo, the extreme reluctance of either the UN or NATO to become involved in ground combat operations led to a sole reliance on air power when the humanitarian crises could no longer be ignored. In each of these cases, the shorter in Bosnia and the longer in Kosovo, even bringing the warring parties to the negotiating table through the precise application of force did nothing to eliminate the underlying tensions. The tensions left unresolved in Bosnia eventually flared again in Kosovo. In the end, foreign troops remained deployed to the region for more than a decade while the nations began the difficult path to stability.

However, while it might seem easy to assign a negative trend based on this observation, it must be remembered that the sole purpose of the bombing was to halt the humanitarian crisis underway and in both cases it did, at least temporarily. The early operations in Afghanistan in 2001–2002, however, highlight the downside that can result from relying heavily on such technology. While the Taliban were quickly removed from power, enabled largely by American air power, they were able to reconstitute themselves as a viable threat requiring a much greater combat effort that continues to the present. As a result, the trend is flat-to-slightly-negative.

Adverse asymmetrical responses Richard Lacquement observed that, "The United States has attained an unmatched mastery of conventional war. This pushes potential challengers to asymmetric responses at both the high and low end of the conflict spectrum and reduces the chance that the United States will become engaged in large-scale conventional war of the style demonstrated during the Persian Gulf War."[22] While he noted that American military mastery "reduces the chance" that the U.S. will fight a major ground war, the invasion of Iraq in 2003 demonstrated that that chance has not yet been reduced to zero.[23] But his comment on "pushing challengers to asymmetric responses" is particularly germane to this study. This line of inquiry has a relatively straight forward metric. Is there evidence of a violent asymmetric response by the enemy that increases levels of human suffering?

While there is no evidence to support a determination along this line in the Vietnam case study, there is an instructive lesson for those who would seek to do battle with the U.S., a lesson which could, unfortunately, encourage an adversary to seek asymmetric alternatives. Towards the end of the war, when North Vietnam transitioned to the third phase of Mao's model, the NVA played directly into the strengths of the American military, suffering significant casualties as a result. In both Iraq and the Balkans, asymmetric moves by the enemy partially blunted the military advantage gained by the U.S.'s use of PGMs.

During DESERT STORM, because he was unable to deal directly with the air assault, Saddam Hussein sought to bring Israel into the war by launching Scud missiles into that country, thereby potentially threatening coalition cohesion. Fortunately, casualties were not particularly high. Later, following the *al Firdos* bunker attack which resulted in high civilian casualties, he announced that Iraqi civilians would subsequently be held at all military bunkers, an announcement which effectively placed them on the coalition no-strike list.[24] Finally, the two incidents involving Kuwait oils fields, the release of oil into the Persian Gulf and the torching of the oil heads, each held the potential for significant human suffering due to ecological disruption although these possibilities were unrealized.

In Bosnia, on at least two instances UN troops were taken hostage, first as a means to embarrass the organization, and second, when physically chained to potential targets, as a means to dissuade the alliance from continuing punitive strikes.[25] In all of these aforementioned cases, while there was a very real potential for increased human suffering by these asymmetric actions, in fact no significant harm was done. Thus, while the summary must acknowledge that these possibilities were very real, the trend is positive. This was not the situation in Kosovo.

As earlier recorded, during ALLIED FORCE Milošević employed a number of indirect, asymmetric means to attempt to disrupt coalition unity and play on the alliance's key concerns about collateral damage and an expanding humanitarian

22 Lacquement Jr., *Shaping*, 160.

23 He did not imply this either. Lacquement's work was written prior to Operation IRAQI FREEDOM.

24 Bush and Scowcroft, *World Transformed*, 454. This was not the first time the issue of "human shields" was raised in Iraq.

25 Beale, *Bombs Over Bosnia*, 25, 29.

crisis. These tactics included: "the use of terror tactics against Kosovar civilians; attempts to exploit the premium the alliance placed on minimizing civilian casualties and collateral damage; creation of enormous refugee flows to create a humanitarian crisis; and, the conduct of disinformation and propaganda campaigns."[26] Because of the heavy reliance on air power, to the exclusion of the introduction of any ground combat forces which may have had at least a chance to counter some of these Serbian tactics, the trend here must be seen as negative.

This trend was reinforced by the experience of American Forces during Operation IRAQI FREEDOM launched in 2003. Relying on the "new" American way of war, in part enabled by PGMs, rapid capitulation by the Iraqis was followed by an almost immediate recognition that there were not enough troops in country to secure the peace, and the multi-year insurgency followed. Until the departure of American troops in late 2011, insurgents rarely sought force-on-force engagements, relying rather on the use of Improvised Explosive Devices (IEDs) and various types of vehicle bombs which caused significant suffering for U.S. troops and Iraqi civilians alike.

Long term suffering The final line of inquiry in the second portion of the analysis focused on the issue of long-term suffering. Underpinning this inquiry is the acknowledgment that discrete combat power, delivered by precision-guided weapons can, while being highly discriminate in the event, cause great human suffering after hostilities are over. As William Arkin correctly observes, "You can't separate neat and clean bombing from postwar deaths, people just died in a different way because of the efficiency of the attacks."[27] These post bellum considerations hold the very real possibility of turning a generally positive determination regarding PGMs effect on human suffering into a strong negative due to suffering experienced in the out years.

As outlined in the first case study, there is little evidence to suggest that the use of PGMs contributed to any significant degree of long-term suffering in Vietnam. Once it had the capability, the U.S. government continued to refrain from destroying the dike system. The impact of the loss of those electrical systems degraded by the Americans was mitigated by "work arounds" the Vietnamese had in place. The issue of unexploded ordnance, an issue in Vietnam and any number of other places, is much more commonly caused by the historically high ratio of unexploded sub-munitions from cluster bombs and uncleared land mines. This is not meant to imply that there aren't any unexploded LGBs buried, albeit deeply, in Vietnamese soil. While the trend is not negative here, it is clear from the evidence that it is in some other cases.

The first of these is DESERT STORM. The story of PGMs is almost universally seen as positive for this conflict, and General Glosson and his "Black Hole" team carefully planned to inflict near-term, but not long-term damage to Iraq's

---

26 Lamb, *Golden Nuggets*, 2–3.

27 Arkin, "Tactical Bombing of Iraqi Forces," 62.

critical infrastructure. But, as outlined earlier, the failure to get that message to all planners in theater resulted in greater damage to Iraq's electrical system than was ever envisioned, resulting in significant post-war suffering.

The 1991 post-war survey conducted by the Harvard University School of Public Health asserted that damage to the electrical system "hampered efforts to purify water and contributed to an alarming escalation in infant mortality rates."[28] Because Iraq relied heavily on electrical power for water purification and distribution, and sewage removal and treatment, the loss of power had significant negative effects.[29] Finding "sharply increased levels of gastroenteritis, cholera, typhoid, and malnutrition in Iraqi children due to the delayed effects of the Gulf War," the study estimated that there were 70,000–90,000 additional deaths, due to increased mortality rates, between April–December 1991.[30] William Arkin estimated that 110,000 additional Iraqis eventually died due to the "indirect detrimental health effects" of the war.[31] These medical conditions show just a portion of the problems caused by the loss of, or reduction in, reliable electrical power.

Arkin also wrote about Operation ALLIED FORCE. What he records is very positive in regard to minimizing collateral damage and long-term suffering, observing that "NATO demonstrated an almost obsessive concern for civilian casualties and damage. ... Enormous efforts were made, largely successful, to minimize short- and long-term civilian effects."[32] Yet, there was for some time a significant disruption to the lives of many living in the region. Finally, even a one-time contingency strike can have significant effects on the population of the country.

The INFINITE REACH strike, while a well-planned and extremely accurate attack, nevertheless inadvertently caused a significant degree of human suffering by eliminating the country's major source of pharmaceuticals. As Werner Daum recorded, the *Al-Shifa* factory was responsible for producing 20 to 60 percent of overall pharmaceuticals in Sudan and, perhaps more critically, 100 percent of the country's intravenous liquids. He noted that it took "more than three months for these products to be replaced with imports."[33] In his opinion, while acknowledging that it is hard to assess, "tens of thousands seems a reasonable guess" to the question of how many Sudanese died due to the destruction of the factory.[34]

As just mentioned, low troop levels in Iraq allowed for the wholesale looting of critical infrastructure following major combat operations. Such looting, that included the widespread removal of copper cabling from the electrical power grid,

28 Gordon and Trainor, *General's War*, 316–17.

29 *Needless Deaths in the Gulf War*, 172.

30 Quoted in Keaney and Cohen, *GWAPS*, Summary Report, 75n.45.

31 Ibid.

32 Arkin, "Operation Allied Force," 28.

33 Daum, "Universalism and the West," 19.

34 Ibid.

merely exacerbated problems with Iraqi infrastructure that had existed since the end of DESERT STORM.

What emerges here is neither completely positive, nor negative. In many of the cases no unintended long-term suffering resulted, which indicates a positive trend. Also, during the period studied the impact of war on the civilian population was far less than in previous conflicts. But such an evaluation must necessarily be tempered by the fact that in a couple of key instances, significant suffering resulted that is clearly a negative trend. Rather than try to somehow quantify the positives and negatives here, it is perhaps best to acknowledge that even positive intentions have the potential for seriously negative consequences—in essence, the Kant *v.* Bentham/Mill conundrum wherein a positive ethical choice as seen from one perspective results in negative outcomes when viewed from the other. The lesson for senior leaders, then, is to openly acknowledge these possibilities are very real, and need to be considered and examined prior to ordering lethal action.

It is also worth addressing Ellul's major concerns regarding *technique* at this point, as they primarily exist at the intersection of the first two levels of analysis. As earlier noted, in his work *The Technological Society*, Ellul was concerned that "technical progress today is no longer conditioned by anything other than its own calculus of efficiency."[35] In an era when technological *means* become the focus, to the relative exclusion of rationally considering *ends*, "certain unforeseen secondary effects ... are much more disastrous than the lack of technique would have been."[36] While the evidence presented makes it clear that the American military is very focused on efficiency, the insidious possibilities of which Ellul was concerned have not yet materialized. As has been demonstrated, at least into the early years of the twenty-first century, political leaders and military planners have been very concerned about, and paid particular attention to, the issues of ends and unintended consequences.

Part III: The Question of Teleology

President George W. Bush, during a campaign speech at the Citadel in 1999, observed that "Those who want to lead America accept two obligations. *One is to use our military power wisely, remembering the costs of war*. The other is to honor our commitments to veterans who have paid those costs."[37] And later, "*Power is increasingly defined, not by mass or size, but by mobility and swiftness*. Influence is measured in information, safety is gained in stealth, *and force is projected on the long arc of precision-guided weapons*. This revolution perfectly matches the strengths of our country—the skill of our people and the superiority of our

---

35 Ellul, *Technological Society*, 74.

36 Ibid., 105.

37 George W. Bush, "A Period of Consequences," Speech (The Citadel, September 23 1999): emphasis added.

technology."[38] These comments were remarkably similar to ones made by his father nearly a decade before.

Then, during remarks made at the Aspen Institute as the crisis in the Gulf initially unfolded, President George H.W. Bush noted,

> Time and again, we have seen technology revolutionize the battlefield. *The U.S. has always relied upon its technological edge to offset the need to match potential adversaries' strength in numbers—cruise missiles, Stealth fighters and bombers, today's 'smart' weapons with the state-of-the-art guidance systems, and tomorrow's 'brilliant' ones*. The men and women in our Armed Forces deserve the best technology America has to offer.[39]

Such remarks highlight the ever greater reliance the U.S. has placed on gaining and maintaining technological advantage. As Lacquement notes,

> New technology and new operational concepts already offer an alternative to the kind of military operation that pits large numbers of young Americans against an adversary in brute, force-on-force conflicts. This new way of war leverages technologically superior U.S. military capabilities to achieve national objectives. It is a strategy of asymmetric force that applies U.S. advantages to strike directly at an adversary's ability to wage war.[40]

Having such an advantage undeniably affects how senior leaders decide when, where, and how to use lethal force in the pursuance of national political objectives. That is the realm of the final part of the evaluative framework, the level that is the least quantifiable and, simultaneously, the most difficult to make a definitive determination. For if our reliance on precision-guided technologies either breeds a reluctance to use more conventional ground forces, thereby creating opportunities for our adversaries, or results in a willingness to use force more often, and if either of these results in greater human suffering, then even the positive conclusions drawn earlier may have to be modified.

Reluctance to deploy conventional land forces This first line of inquiry is concerned with whether the availability of PGMs breeds a reluctance to deploy ground forces, which ultimately results in greater human suffering. The evidence may signal the first beginning of a significant shift in American strategic thought, although there should be more study of this. While the seven contingency operations offer no insight here, the other four conflicts studied have led to markedly different conclusions.

38 Ibid.: emphasis added.

39 Bush, *Aspen Institute Symposium*: emphasis added.

40 Lacquement Jr., *Shaping*, 57.

In the Vietnam case, it was clear that the U.S. was willing to deploy ground forces into most parts of the theater. The decision not to move ground forces into North Vietnam rested solely on the unwillingness of the Johnson Administration to risk becoming involved in a wider conflict with the USSR or China. Even after LGBs became operational in 1967, there is no convincing evidence of a subsequent reluctance to put ground troops in harm's way, although Littauer and Uphoff believe that "the Laos campaign ... [was] a clear application of that aspect of the Nixon Doctrine which rest[ed] upon *the substitution of American technology*, in conjunction with Asian forces, *for American manpower*."[41]

Similarly, despite considerable discussion about the utility and effectiveness of air power in the period before Operation DESERT STORM, the Pentagon and Bush Administration resisted the seductive "air only" option, choosing instead to deploy more than 540,000 troops to the theater prior to the ground war. These two cases would represent a positive trend according to the framework. However, the success of this latter operation, coupled with continued improvements in PGM technology, may have begun an inexorable swing in the other direction.

While clear that they did not want to get involved, when the Europeans and Americans finally realized they had to act in the Balkans they relied almost exclusively on air power to bring the warring parties to the negotiating table. In Bosnia, as earlier recorded, even though there were some UN troops on the ground, the reluctance to deploy viable combat forces in theater, and to instead rely on the odd NATO air strike to coerce the Serbs, was to have a significant impact on the Muslim population of Srebrenica. The International Committee of the Red Cross would later determine that just over 7,000 Muslims were slaughtered between 12 and 16 July, most dying in ambushes and mass executions.[42] A similar situation took place in Kosovo

President Clinton's publicly stated position of "no ground forces" had significant consequences for Albanians in the region. Some of the greatest massacres by Serbian forces, in Cuska, Podjevo, and Velika Krusa, occurred during the air campaign. According to one estimate made by the British government, about 10,000 ethnic Albanians were killed during ALLIED FORCE.[43] Also, within just the first few days of the air campaign thousands of Kosovar Albanians began fleeing their homes in what one author called "a ghastly exodus that would quickly reach near biblical proportions."[44] By the end of the 78-day campaign, the UNHCR calculated that more than 848,000 people fled the province, roughly 40 percent of the population.[45]

---

41 Littauer and Uphoff, eds., *Air War in Indochina*, 80: emphasis added.

42 Cited by Holbrooke, *To End a War*, 70. Another number often seen in accounts is 8,000.

43 Lambeth, *NATO's Air War*, 60.

44 Nation, *War in the Balkans*, 249. Lambeth points out that during the early part of the air war, most of 1.8 million ethnic Albanians were driven from their homes: Lambeth, *NATO's Air War*, 26.

45 Nation, *War in the Balkans*, 249–50.

These two examples clearly exhibit negative trends. More recent operations, however, only partially support this conclusion. In the cases of Afghanistan and Iraq (2003), national leaders were willing to commit troops, indeed increase troop strengths when a surge was deemed necessary, in an effort to end the conflicts. The public announcement by President Barak Obama in 2011, that no U.S. ground troops would be sent into Libya trends the other way.

Greater willingness to resort to lethal force The final line of inquiry sought an answer to the question of whether having these "tools" at their disposal encourages leaders to use violence more frequently, before traditional just war considerations have been satisfied or even addressed. If true, then a technology that reduces human suffering due to its inherent accuracy and ability to be used discriminately could, paradoxically, be viewed in a negative light. A review of some commentators indicates that such a trend may in fact be underway.

In the preface to the Cornell University study, Neil Sheehan asks the question, "[H]ow [was] the current Administration of President Richard M. Nixon ... able to continue an unpopular war in Indochina for four years after the country had voted against that war in the presidential election of 1968?"[46] His answer is enlightening. He believes, in large measure, that "technological advances in weaponry, chiefly in air power" are the answer.[47] Specifically:

> The development of the air weapon to its present responsiveness to centralized control, its remoteness from the tides of the ground battle, and its vast destructive force have created the opportunity for an American President and the state machinery that serves him in the Executive Branch to conduct war with little reference to the wishes of the body politic at home. The reason is simple. War from the air is not very tangible to the average American.[48]

While he doesn't specifically address the issue of LGBs in his comments, their availability, demonstrated accuracy, and reliability were key factors in enabling President Nixon to prosecute targets near Hanoi and Haiphong during the two LINEBACKER operations.

White extends this argument believing that President Ronald Reagan heralded a new era when, in March 1986, he declared, "the United States would respond to any threats to its national interests by 'strik[ing] back with exquisite calibration on a schedule of its own choosing, and in a way that presses its advantages in economic power and military technology, [and] retaining popular support at home by avoiding as much as possible the expenditure of U.S. lives.'"[49]

---

46 Neil Sheehan, preface to *Air War in Indochina*, Littauer and Uphoff, eds., v.

47 Ibid.

48 Ibid.

49 Quoted in, White, *Crises After the Storm*, 8.

SECDEF Cheney, in his first post-DESERT STORM annual report to the president and congress stated that "tactical air forces 'constitute a powerful and highly flexible component of the U.S. deterrent' and, as such, 'are an essential element of this nation's crisis response capability."[50] Benjamin Lambeth writes of "cruise missile diplomacy," while "Major General Charles Wald, USAF, in a speech given in 1998, declared that 'Air power has become the instrument of choice in America's foreign policy. Air forces, because they are extremely accurate and less vulnerable than ground forces, provide a politically viable instrument of force and oversight to national authorities.'"[51] It is clear that PGMs enable the use of force more often, with significantly less risk of a major negative outcome.

But it was also evident that, at least through 1999, senior decision-makers did address the ethical and legal bases for the use of force prior to ordering action. While critics may disagree as to whether such justifications such as Article 51 of the UN Charter are sufficient, to that point lethal force had not been employed in a random or unconsidered fashion. Additionally, as provided for in the framework, in at least two cases, Libya and Iraq (1993), evidence suggests that the early use of force dampened future actions on the part of those regimes, and that helps dampen a negative trend.

Conclusions

Do PGMs mitigate suffering in war and have these weapons changed the way decisions regarding war and peace have been made? For the period 1967–1999, the answers to both parts of this underlying question are relatively clear and straightforward. However, it is also clear that there appear to be trends emerging that mean this question might be answered much differently in the future. From all the evidence obtained and analyzed during this study, it is evident that PGMs have indeed had a significant, positive influence on reducing human suffering in war. While, again, not seeking to minimize the suffering of anyone affected by the primordial violence of war, conflicts today are far less lethal endeavors than in the past. The "American Way of War" described by Russell Weigley in his landmark studyof the same title, massive build-ups of forces with the intent of achieving a crushing victory over the enemy, has given way to the modern, precise warfare of today's American military. There are two key reasons for this; precision and a commitment to reducing civilian casualties and collateral damage

While there are several key elements to the MTR/RMA phenomenon experienced by the U.S. military in the last four decades, high-speed computing and greater surveillance capability among them, precision remains the key enabler

50 Quoted in, Gillespie, *Precision Guided Munitions*, 222.

51 Quoted in, White, *Crises After the Storm*, 83. The speech was given at the Center of Strategic and International Studies symposium, "Dueling Doctrines: The New American Way of War," Washington, June 25, 1998.

as evidenced by the president's reference in the epigraph at the beginning of this chapter, among many others. Where once tens of thousands of civilians might be killed in a single bombing raid, because the only way to influence the enemy at that time was to raze their cities, or so it was believed, today precision allows the U.S. to target key components of an adversary's infrastructure in an attempt to achieve a similar result.

This is possible because for the three decades studied here, PGMs have been extremely accurate and highly reliable, particularly when compared to the aerial weapons of World War II, Korea, and the early part of Vietnam. When employed correctly by a well-trained crew, in the appropriate meteorological conditions for a given weapon, the vast majority of the time that weapon, if not a direct hit, will hit within a few paces of the exact point the planners wanted it to impact.

This technological capability, demonstrated across diverse operational environments and with the various types of PGMs in the American weapons inventory, has been coupled with an explicitly stated desire to try to reduce civilian casualties and long-term damage to infrastructure to the greatest extent possible. From the evidence presented, primarily in the first level of analysis, between 1967 and 1999 human suffering has been significantly reduced in operations involving American forces and PGMs. Where, in the past, many thousands of people died, now sometimes only tens perish in an attack, a trend that continues to the present.

But it must also be acknowledged that, as seen in the study, however well designed the weapons, however well trained the crews, and however well-intentioned the planners and leaders, everybody involved with this process is human and errors will be made. In fact, Ellul acknowledges that such human error will occur. These results were revealed in the second level analysis. Targets can still be misidentified, particularly under the stress of lethal anti-aircraft fire as in Libya (1986). Commander's guidance might not be promulgated to all appropriate personnel as in Iraq (1991). Intelligence, as was seen in the Balkans and likely in the Sudan, can be wrong. Similarly, any technology has some possibility of failure. Things such as a rocket motor misfiring, a seeker-head's gimbals breaking, a guidance fin failing to fully deploy, or any number of complex micro components not operating as designed are rare but real possibilities. For example, friendly forces were killed in Afghanistan by an airstrike due to a little known problem with a GPS system resetting coordinates following a battery change.

Events such as these can and likely will result in human suffering, as demonstrated by events that occurred during DESERT STORM, the Balkans, and after INFINITE REACH in Sudan, caused either by the weapons directly or by second or third order effects. However, despite the instances described earlier in the narrative, the overall trend is still predominantly positive. It is at this juncture, as the analysis moves from the second to the third level of analysis, that the analysis becomes a bit more complicated.

Past success breeds future expectations. This "New American Way of War," enabled in large part by precision, has implanted a number of expectations in the minds of some political leaders, some in the military, and the American public. It is broadly expected that combat operations will always be clean, that there will be relatively little bloodshed (particularly on the part of the American forces), and that there will be little collateral damage. Such capabilities and expectations undoubtedly influence key decision-makers. Possessing the capability to replace massed forces with focused firepower, and with the accompanying expectation of few casualties, should not lead to a more rapid application of lethal force, however seductive that option might appear in a given crisis situation. NATO's recent engagement in Libya may provide insight in the future.

However, as with any other military operation being contemplated, if all aspects have been carefully considered, the actions are deemed in the national interest, necessary, and justifiable, then it should be seen as positive that an administration would seek to use the least amount of combat power necessary. Such decisions are in keeping with the centuries old Just War concept of proportionality in combat. Lower levels of firepower properly applied should, ideally, result in the attainment of the desired objective with lower levels of human suffering. This type of victory, a "Quick Decisive Victory" or QDV in U.S. military parlance, which combines a quick military victory with the rapid attainment of the desired political objective, is just the type of operation Americans have come to expect.

These self-reinforcing expectations may explain why, at times, cruise missile strikes have been launched in situations that were hard to understand by external observers. One immediately thinks of Operations DESERT STRIKE and INFINITE REACH in this context. So far, however, nothing has been identified in this area that would, at the third level of analysis, alter the positive conclusions. But what of the future?

Potentially Troubling Trends?

From observable trends and current political rhetoric, the United States is unlikely to walk away from its comparative military advantage any time soon. Innovations in the field of precision-guided weapons continue. Smaller precision-guided munitions, the GBU-39/B Small Diameter Bomb (SDB) for example, have been developed to engage targets with lower collateral damage.[52] "Smart weapons" are scheduled to be replaced, in some cases, by "brilliant weapons," those capable of making independent targeting decisions. There has also been discussion of replacing nuclear warheads on some Inter-continental Ballistic Missiles (ICBMs) with conventional warheads to provide a rapid response, long-range, precision strike capability to the national leadership as part of "Prompt Global Strike" initiative. The military has dropped inert, concrete-filled training ordnance into a densely populated area in Iraq to destroy a target building without placing any surrounding structures, or their inhabitants, at risk.

---

52 The aircraft can also carry more of these smaller weapons than it could with the larger 1,000 or 2,000-pound weapons.

However, trends have been emerging over the last decade that are a cause for concern and should give pause to ethicists, political and military leaders, and those they serve. The first is the issue of PGM-armed Unmanned Aerial Vehicles (UAVs), also commonly referred to as "armed drones." For almost a decade, the U.S. has used armed UAVs to strike enemy combatants, in areas within *and* beyond active combat zones.[53] An early example, the attack on senior Al Qaeda operative Abu Ali al-Harithi in Yemen in November 2002, garnered much press coverage and was the subject of a featured case study in the *Journal of Military Ethics* in 2003.[54] Until very recently, unless there was a "friendly fire" incident, or claim of civilian deaths, the use of UAVs had become so commonplace that they seemed not worthy of note in the popular press, although the attack that killed Anwar al-Awlaki, a U.S. citizen, did ignite debate in the national media.[55] However, as this book was going to press in the summer of 2012, that public silence appeared to be giving way as more and more voices were raising concern over the legal and ethical issues related to the use of such drones.

This does not imply that all had remained silent previously. Peter Singer's most recent book, *Wired for War: The Robotics Revolution and Conflict in the 21st Century*, documents the trend towards robotic warfare and raises concerns that this will "lower the moral and psychological barriers to killing."[56] He is joined by others such as Armin Krishnan, *Killer Robots: Legality and Ethicality of Autonomous Weapons* (2009), Ronald Arkin, *Governing Lethal Behavior in Autonomous Robots* (2009), and Noel Sharkey whose essay "Saying 'No!' to Lethal Autonomous Targeting" was a featured part of a 2010 special issue of the *Journal of Military Ethics* devoted to "Ethics and Emerging Military Technologies."[57]

There are a least three related points of concern. The first is that, until recently, the public had seemingly become numb to the use and expansion of unmanned combat and surveillance operations around the world. This may have been a function of a nation having been at war for more than a decade, a case of a very short attention span in an information-filled world, or a concrete example of Ellul's concerns expressed in *The Humiliation of the Word*. Little comment was made, for example, to revelations in the fall of 2011 that secret drone bases were being operated, or being developed, in Djibouti, the Seychelles, and Ethiopia. While these drone might not all be armed, they certainly do have that capability if the need arises. While secret bases are certainly not a new phenomenon, these can be/

53 I acknowledge that others have argued that, in this war on "terror," formerly clear distinctions such combat zones and non-combat zones are no longer valid.

54 See *Journal of Military Ethics* 2.3 (2003).

55 I will not attempt to resolve the debate of whether Anwar al-Awlaki was a legitimate military target.

56 Peter W. Singer, *Wired for War: The Robotics Revolution and Conflict in the 21st Century* (New York: Penguin Press, 2009), flyleaf.

57 Details can be found in the Bibliography.

have been used to launch missions into nations, such as Somalia and Yemen, that are not combat zones in the traditional sense.

The second concern revolves around the greater willingness of senior political and military leaders to use drones as offensive weapons. If the public regards these operations as commonplace, then the decision to use armed UAVs potentially becomes easier and less deeply considered. Simply stated, past success can breed future expectations, expectations that can also influence future decision-making. If decision-makers accept precision's near infallibility, and possess the capability to employ PGMs without the risk of the loss of aircrew, then it is entirely conceivable that a leader's internal threshold for ordering combat operations will be lowered below the point where, in the past, *jus ad bellum* considerations would have argued against lethal action.

The third concern regarding UAVs involves unintended consequences. Here, three aspects are present. The first is the fact that by using standoff strike capabilities the U.S. effectively cedes the post-strike public relations initiative to the adversary. Claims of unnecessary civilian casualties, including those of local wedding parties, have closely followed several strikes into Afghanistan. Whether these claims were ultimately found to be true or not, recall the false reporting out of North Vietnam earlier, the initial wave of negative press reports in this "instant news" world had to be dealt with and could become a distraction for senior political and military leaders. Corrections to the initial story, published some weeks later, carry much less weight.

The second is related to the first in that using PGMs can breed two separate emotions among the local populace. The first is rage and a sense of a desire for revenge. Those subject to standoff strikes have no immediate recourse; it is not heroic warfare in the traditional sense. Couple that frustration with press reporting and popular sentiment and it must be wondered if future adversaries are being conditioned by these actions. The other emotion, evident in Iraq during 1990–1991, is that of "so what?"

If, in the minds of the people, there is little personal fear of war, and no attendant feeling of being "defeated" when the war is over, then there would seem to be less of a reluctance to engage in a future struggle. If, as Clausewitz believed, the results of any given conflict are never final, that the vanquished merely waits for the right opportunity to reverse the results in the future, then there may well be a paradoxical danger in being too precise. Such a dynamic may also exist in the minds of the leaders. An unrepentant Saddam Hussein shortly after the 1991 cease-fire was declared may be an early example, as was an unbowed Slobodan Milošević later in the Balkans.

Finally, and this has been most evident with recent U.S.–Pakistan relations, there is the political aspect. The ability and willingness to use drones with relative impunity into neutral and allied nations brings the potential for diplomatic tensions and fallout. At least twice during 2011, during the spring and fall, Pakistan demanded a cessation of drone strikes on its territory. The knock-on effect in the fall instance was the shutting down of critical supply lines into Afghanistan, a route that remains closed as this is being written.

A second trend, less certain at this juncture, is whether possessing the ability to precisely strike just about any target is beginning to transform the concept of "war" in the minds of leaders. One example would be the willingness to expand lethal actions into non-war zones nations like Somalia, Pakistan, and Yemen. But perhaps more insightful is the Obama administration's public discourse on the recent action to overthrow the Libyan regime. While American aircrews flew combat and combat support missions, and missiles were launched against targets in Libya, the administration refused to characterize this operation as "the kind of 'hostilities' contemplated by the [War Powers] Resolution.[58] If, as was adopted earlier in this volume, war is "an act of force to compel our enemy to do our will," then it is hard to see how this self-described "military action" does not meet this criterion.[59]

In the final analysis, the ongoing use of PGMs as a tool of warfare stands at an interesting crossroad. Other technologies, such as gunpowder and artillery, earlier held the promise of enabling lower human suffering in war. These hopes were, unfortunately, unrealized. This study has determined that for the last several decades, PGMs have earned this reputation. However, as potentially troubling trends emerge it may well be wise to remember that while the Kantian "ought implies can" is an accepted ethical tenet, the corollary remains true; that we can *does not* mean we should. Ensuring that established Just War considerations remain part of the debates regarding the use of force, particularly as technology makes it ever easier to employ weapons with little or no risk to our own forces, will help keep the moral component alive in this martial pursuit.

58 Report to Congress on United States Activities in Libya. 15 June 2011. p. 25. Available at: https://www.documentcloud.org/documents/204732-united-states-activities-in-libya-6-15-11.html (accessed 15 December 2011).

59 Carl Von Clausewitz, *On War*, ed. and trans. Michael Howard and Peter Paret (Princeton, NJ: Princeton University Press, 1976), 75.

Selected Bibliography

Books/Chapters in Books/Monographs

Akhavan, Payam, and Robert Howse eds. *Yugoslavia the Former and the Future: Reflections by Scholars from the Region*. Washington, DC: Brookings Institution, 1995.

Albright, Madeleine K. (Korbel) with Bill Woodward. *Madam Secretary: A Memoir*. New York: Miramax Books, 2003.

Amstutz, Mark R. *International Ethics: Concepts, Theories, and Cases in Global Politics*. New York: Rowman & Littlefield, 1999.

Anderegg, C.R. *Sierra Hotel: Flying Air Force Fighters in the Decade After Vietnam*. Washington, DC: U.S. Government Printing Office, 2001.

Anderson, Benedict. *Imagined Communities: Reflections on the Origin and Spread of Nationalism*. Revised. New York: Verso, 1991.

Aristotle: I. Great Books of the Western World, ed. Robert Maynard Hutchins, vol. 8. Chicago: Encyclopedia Britannica, 1952.

Aristotle: II. Great Books of the Western World, ed. Robert Maynard Hutchins, vol. 9. Chicago: Encyclopedia Britannica, 1952.

Arkin, Ronald C. *Governing Lethal Behavior in Autonomous Robots*. Boca Raton, FL: CRC Press, 2009.

Arkin, William. "Operation Allied Force: 'The Most Precise Application of Air Power in History.'" In *War Over Kosovo: Politics and Strategy in a Global Age*, ed. Andrew J. Bacevich and Eliot A. Cohen, 1–37. New York: Columbia University Press, 2001.

Art, Robert J., and Kenneth N. Waltz. "Technology, Strategy, and the Uses of Force." In *The Use of Force: International Politics and Foreign Policy*, ed. Robert J. Art and Kenneth N. Waltz, 1–32. New York: University Press of America, 1983.

Babbie, Earl. *The Practice of Social Research*. 9th ed. Belmont, CA: Wadsworth, 2001.

Bacevich, Andrew J., and Eliot A. Cohen eds. *War Over Kosovo: Politics and Strategy in a Global Age*. New York: Columbia University Press, 2001.

Barbour, Ian G. *Ethics in an Age of Technology: The Gifford Lectures Volume Two*. San Francisco: HarperSanFrancisco, 1993.

Benda, Julien. *The Living Thoughts of Kant*. ed. Alfred O. Mendel. The Living Thoughts Library. Philadelphia: David McKay Co., 1940.

Bentham, Jeremy. "An Introduction to the Principles of Morals and Legislation." In *The Utilitarians*, 5–398. Garden City, New York: Dolphin Books, 1961.

Black, Jeremy, ed. *European Warfare 1815–2000*. Problems in Focus. New York: Palgrave, 2002.

Blackwelder, Donald I. *The Long Road to Desert Storm and Beyond: The Development of Precision Guided Bombs*. Maxwell AFB, AL: Air University Press, 1993.

Brennan, Joseph Gerard. *Foundations of Moral Obligation: The Stockdale Course*. Newport: Naval War College Press, 1992.

Brigety II, Reuben E. *Ethics, Technology and the American Way of War: Cruise Missiles and US Security Policy*. Contemporary Security Studies. New York: Routledge, 2007.

Brodie, Bernard. "Technical Changes, Strategic Doctrine, and Political Outcomes." In *Historical Dimensions of National Security Problems*, ed. Klaus Knorr. Lawrence: University Press of Kansas, 1976.

——. *War and Politics*. New York: Macmillan Publishing Co., 1973.

Bush, George (Herbert Walker), and Brent Scowcroft. *A World Transformed*. New York: Vintage Books, 1998.

Cameron, Gavin. *Nuclear Terrorism: A Threat Assessment for the 21st Century*. London: Macmillan Press Ltd., 1999.

Campbell, Christopher M. "The Deliberate Force Air Campaign Plan." In *Deliberate Force: A Case Study in Effective Air Campaigning: Final Report of the Air University Balkans Air Campaign Study*, ed. Robert C. Owen, 87–130. Maxwell AFB, AL: Air University Press, 2000.

Capra, Fritjof. *The Turning Point: Science, Society, and the Rising Culture*. New York: Bantam, 1982.

Carus, W. Seth. *Cruise Missile Proliferation in the 1990s*. The Washington Papers, ed. The Center for Strategic and International Studies, vol. 159. Westport, CT: Praeger, 1992.

Chin, Warren. "The Transformation of War in Europe 1945–2000." In *European Warfare 1815–2000*, ed. Jeremy Black, 192–217. New York: Palgrave, 2002.

Christopher, Paul Pasquale. *The Ethics of War and Peace: An Introduction to Legal and Moral Issues*. 3rd. ed. Upper Saddle River, NJ: Pearson Prentice Hall, 2004.

Christopher, Warren. *In the Stream of History: Shaping Foreign Policy for a New Era*. Stanford, CA: Stanford University Press, 1998.

Cipolla, Carlo. *Guns, Sails, and Empires: Technological Innovation and the Early Phases of European Expansion, 1400–1700*. New York: Pantheon Books, 1965.

Clark, Grahame. "The Beginning of Farming: Jericho and Jarmo." In *The World of the Past*, ed. Jacquetta Hawkes, 249–53. New York: Alfred A. Knopf, 1963.

Clark, Wesley K. *Waging Modern War: Bosnia, Kosovo, and the Future of Conflict*. New York: Public Affairs, 2001.

Clausewitz, Carl Von. *On War*. Edited and translated by Michael Howard and Peter Paret. Princeton, NJ: Princeton University Press, 1976.

Clodfelter, Mark. *The Limits of Airpower: The American Bombing of North Vietnam*. New York: The Free Press, 1989.

Collins, Joseph J. *Choosing War: The Decision to Invade Iraq and Its Aftermath*. Institute for National Strategic Studies: Occasional Paper, vol. 5. Washington, DC: National Defense University Press, 2008.

Conflict and Cooperation: Evolving Theories of International Relations. ed. Mark A. Genest. 2nd. ed. Belmont, CA: Thomson Wadsworth, 2004.

Conversino, Mark J. "Executing Deliberate Force, 30 August–14 September 1995." In *Deliberate Force: A Case Study in Effective Air Campaigning: Final Report of the Air University Balkans Air Campaign Study*, ed. Robert C. Owen, 131–76. Maxwell AFB, AL: Air University Press, 2000.

Cooper, Chester L. *The Lost Crusade: America in Vietnam*. New York: Dodd, Mead & Co., 1970.

Cordesman, Anthony H. *The Lessons and Non-Lessons of the Air and Missile Campaign in Kosovo*. Westport, CT: Praeger, 2001.

Davis, Bradley S. "The Planning Background." In *Deliberate Force: A Case Study in Effective Air Campaigning: Final Report of the Air University Balkans Air Campaign Study*, ed. Robert C. Owen, 37–63. Maxwell AFB, AL: Air University Press, 2000.

Deitchman, Seymour. *Military Power and the Advance of Technology: General Purpose Military Forces for the 1980s and Beyond*. Boulder, CO: Westview Press, 1979.

DeLeon, Peter. *The Laser-Guided Bomb: Case History of a Development*. Rand no. 1312-1. Santa Monica, CA, 1974.

Detter, Ingrid. *The Law of War*. 2nd ed. Cambridge, UK: Cambridge University Press, 2000.

Dickson, Paul. *The Electronic Battlefield*. Bloomington: Indiana University Press, 1976.

Digby, James F. *Precision-Guided Weapons: New Chances to Deal with Old Dangers*. Rand no. 5384. Santa Monica, CA, 1975.

Downes, Alexander B. *Targeting Civilians in War*. Cornell Studies in Security Affairs. Ithaca, NY: Cornell University Press, 2008.

Ellul, Jacques. *The Humiliation of the Word*. Translated by Joyce Main Hanks. Grand Rapids, MI: William B. Eerdmans Publishing Co., 1985.

——. *The Technological Society*. Translated by John Wilkinson. New York: Vintage Books, 1964.

Ellul, Jacques, and Edward C. Mann III. *Thunder and Lightning: Desert Storm and the Airpower Debates*. Maxwell AFB, AL: Air University Press, 1995.

Ellul, Jacques, and Patrick Troude-Chastenet. *On Religion, Technology, and Politics: Conversations with Patrick Troude-Chastenet*. Translated by Joan Mendes France. South Florida-Rochester-Saint Louis Studies on Religion and the Social Order, vol. 24. Atlanta: Scholars Press, 1998.

Elshtain, Jean Bethke. *Just War Against Terror: The Burden of American Power in a Violent World*. New York: Basic Books, 2003.

Ethics and Foreign Policy. ed. Karen E. Smith and Margot Light. Cambridge: Cambridge University Press, 2001.

Fasching, Darrell J. "From Genocide to Global Ethics by Way of Storytelling." In *Explorations in Global Ethics: Comparative Religious Ethics and Interreligious Dialogue*, ed. Sumner B. Twiss and Bruce Grelle, 291–321. Boulder, CO: Westview Press, 2000.

Feenberg, Andrew. *Alternative Modernity: The Technological Turn in Philosophy and Social Theory*. Berkeley: University of California Press, 1995.

——. "Escaping the Iron Cage: Subversive Rationalization and Democratic Theory." In *Democratising Technology. Ethics, Risk and Public Debate*, ed. Rene Schomberg, 1–15. Tilburg: International Centre for Human and Public Affairs, 1998.

——. *Questioning Technology*. New York: Routledge, 1999.

——. "Subversive Rationalization: Technology, Power, and Democracy." In *Technology and the Politics of Knowledge*, ed. Andrew Feenberg and Alastair Hannay. Bloomington: Indiana University Press, 1995.

Freedman, Lawrence, and Efraim Karsh. *The Gulf Conflict 1990–1991: Diplomacy and War in the New World Order*. Princeton, NJ: Princeton University Press, 1993.

Friedman, Norman. *Desert Victory: The War for Kuwait*. Annapolis: Naval Institute Press, 1991.

Gordon, Michael R., and General Bernard E. Trainor. *COBRA II: The Inside Story of the Invasion and Occupation of Iraq*. New York: Pantheon Books, 2006.

——. *The General's War: The Inside Story of the Conflict in the Gulf*. New York: Little, Brown and Company, 1995.

Gow, James. *Triumph of the Lack of Will: International Diplomacy and the Yugoslav War*. New York: Columbia University Press, 1997.

Handel, Michael I. *Masters of War: Classical Strategic Thought*. 3rd. ed. London: Frank Cass, 2001.

Harris, Elizabeth J. "Buddhism and the Justification of War: A Case Study from Sri Lanka." In *Just War in Comparative Perspective*, ed. Paul Robinson, 93–108. Burlington, VT: Ashgate, 2003.

Harrison, John B., and Richard E. Sullivan. *A Short History of Western Civilization*. 4th ed. New York: Alfred A. Knopf, 1975.

Hazelton, Fran, ed. *Iraq Since the Gulf War: Prospects for Democracy*. London: Zed Books Ltd., 1994.

Hendrickson, Ryan C. *Diplomacy and War at NATO: The Secretary General and Military Action After the Cold War*. Columbia, MO: University of Missouri Press, 2006.

Henriksen, Dag. *NATO's Gamble: Combining Diplomacy and Airpower in the Kosovo Crisis 1998–1999*. Annapolis: Naval Institute Press, 2007.

Herring, George C. *LBJ and Vietnam: A Different Kind of War*. An Administrative History of the Johnson Presidency Series. Austin: University of Texas Press, 1994.

Hinman, Lawrence M. *Ethics: A Pluralistic Approach to Moral Theory*. 2nd. ed. New York: Thomson Wadsworth, 1998.

Hiro, Dilip. *Desert Shield to Desert Storm: The Second Gulf War*. London: Paladin, 1992.

Holbrooke, Richard. *To End a War*. New York: Random House, 1998.

Hollenbach, David. "Ethics in Distress: Can There Be Just Wars in the Nuclear Age?" In *The Nuclear Dilemma and the Just War Tradition*, ed. William V. O'Brien and John Langan, 13–29. Lexington, MA: Lexington Books, 1986.

Huchthausen, Peter A. *America's Splendid Little Wars: A Short History of U.S. Military Engagements: 1975–2000*. New York: Viking Press, 2003.

Hughes, Gerald J. *Routledge Philosophy Guidebook to Aristotle on Ethics*. Routledge Philosophy Guidebooks, ed. Tim and Jonathan Wolff Crane. London: Routledge, 2001.

Human Rights Watch/Helsinki. *Bosnia-Hercegovina: The Fall of Srebrenica and the Failure of U.N. Peacekeeping*. Tech. Report. Vol. 7, No. 13. New York: Human Rights Watch/Helsinki, 1995.

Ignatieff, Michael. *Virtual War: Kosovo and Beyond*. New York: Metropolitan Books, 2000. *International Politics: Enduring Concepts and Contemporary Issues*. ed. Robert J. Art and Robert Jervis, 4th ed. New York: HarperCollins, 1996.

Janjić, Dušan. "Resurgence of Ethnic Conflict in Yugoslavia: The Demise of Communism and the Rise of the 'New Elites' of Nationalism." In *Yugoslavia the Former and the Future: Reflections by Scholars from the Region*, ed. Payam Akhavan and Robert Howse, 29–44. Washington, DC: Brookings Institution, 1995.

Jasani, Bhupendra, ed. *Outer Space: A New Dimension of the Arms Race*. London: Taylor & Francis, 1982.

Jefferson, Thomas. *The Political Writings of Thomas Jefferson: Representative Selections*. ed. Edward Dumbauld. The American Heritage Series, vol. 9. New York: The Liberal Arts Press, 1956.

Jervis, Robert. *The Meaning of the Nuclear Revolution: Statecraft and the Process of Armageddon*. Ithaca, NY: Cornell University Press, 1989.

Johnson, James T. (Turner) *Morality and Contemporary Warfare*. New Haven: Yale University Press, 1999.

——. *The War to Oust Saddam Hussein: Just War and the New Face of Conflict*. New York: Rowman & Littlefield, 2005.

Johnson, James T. (Turner), and George Weigel. *Just War and the Gulf War*. Washington, DC: Ethics and Public Policy Center, 1991.

Jones, John D., and Marc F. Griesbach, eds. *Just War Theory in the Nuclear Age*. New York: University Press of America, 1985.

Kane, Thomas M. "Inauspicious Tools: Chinese Thought on the Morality of Warfare." In *Just War in Comparative Perspective*, ed. Paul Robinson, 139–52. Burlington, VT: Ashgate, 2003.

Kant. Great Books of the Western World, ed. Robert Maynard Hutchins, vol. 42. Chicago: Encyclopedia Britannica, 1952.

Kant, Immanuel. *Foundations of the Metaphysics of Morals: With Critical Essays*. Translated by Lewis White Beck. Edited by Robert Paul Wolff. The Bobbs-Merrill Text and Commentary Series. Indianapolis: Bobbs-Merrill Educational Publishing, 1969.

Keaney, Thomas A., and Eliot A. Cohen. *Revolution in Warfare?: Air Power in the Persian Gulf*. Annapolis: Naval Institute Press, 1995.

Keegan, John. *The Face of Battle*. New York: Viking Press, 1976.

Kelsay, John. "Implications of Islamic Political Thought." In *Explorations in Global Ethics: Comparative Religious Ethics and Interreligious Dialogue*, ed. Sumner B. Twiss and Bruce Grelle, 217–36. Boulder, CO: Westview Press, 2000.

King, Robert H. *Thomas Merton and Thich Nhat Hanh: Engaged Spirituality in an Age of Globalization*. New York: Continuum, 2001.

King, Sallie. "A Global Ethic in the Light of Comparative Religious Ethics." In *Explorations in Global Ethics: Comparative Religious Ethics and Interreligious Dialogue*, ed. Sumner B. Twiss and Bruce Grelle, 118–40. Boulder, CO: Westview Press, 2000.

Knorr, Klaus. *Military Power and Potential*. Lexington, MA: Heath, 1970.

Krishnan, Armin. *Killer Robots: Legality and Ethicality of Autonomous Weapons*. Surrey: Ashgate Publishing, 2009.

Lacquement Jr., Richard A. *Shaping American Military Capabilities After the Cold War*. Westport, CT: Praeger, 2003.

Lamb Sr., Michael W. *Operation Allied Force: Golden Nuggets for Future Campaigns*. The Maxwell Papers, vol. 27. Maxwell AFB, AL: Air University Press, 2002.

Lambeth, Benjamin S. *NATO's Air War for Kosovo: A Strategic and Operational Assessment*. Rand no. 1365. Santa Monica, CA, 2001.

Lavine, T.Z. *From Socrates to Sartre: The Philosophical Quest*. New York: Bantam, 1984.

Lehman Jr., John F. *Command of the Seas*. New York: Charles Scribner's Sons, 1988.

Lewy, Guenter. *America in Vietnam*. New York: Oxford University Press, 1978.

Liotta, P.(Peter) H. *Dismembering the State: The Death of Yugoslavia and Why It Matters*. New York: Lexington Books, 2001.

Littauer, Raphael, and Norman Uphoff, eds. *The Air War in Indochina*. Revised. Boston: Beacon Press, 1972.

Magaš, Branka. *The Destruction of Yugoslavia: Tracking the Break-up 1980–92*. New York: Verso, 1993.

Mahnken, Thomas G. *Technology and the American Way of War Since 1945*. New York: Columbia University Press, 2008.

McCarthy, James R., and George B. Allison. *Linebacker II: A View from the Rock*. USAF Southeast Asia Monograph Series, vol. IV, Monograph 8. Maxwell AFB, AL: Air War College, 1979.

McLaughlin, Mark C. "Assessing the Effectiveness of Deliberate Force: Harnessing the Political-Military Connection." In *Deliberate Force: A Case Study in Effective Air Campaigning: Final Report of the Air University Balkans Air Campaign Study*, ed. Robert C. Owen, 189–98. Maxwell AFB, AL: Air University Press, 2000.

McNamara, Robert S., with Brian VanDeMark. *In Retrospect: The Tragedy and Lessons of Vietnam*. New York: Time Books, 1995.

McNeill, William H. *The Pursuit of Power: Technology, Armed Force, and Society Since A.D. 1000*. Chicago: University of Chicago Press, 1982.

Mill, John Stuart. "On Liberty." In *The Utilitarians*, 473–600. Garden City, New York: Dolphin Books, 1961.

——. "Utilitarianism." In *The Utilitarians*, 399–472. Garden City, New York: Dolphin Books, 1961.

Mitcham, Carl. *Thinking Through Technology: The Path Between Engineering and Philosophy*. Chicago: University of Chicago Press, 1994.

Momyer, William W. *Air Power in Three Wars*. 1979. Reprint. Maxwell AFB, AL: Air University Press, 2003.

Morganthau, Hans J. *Politics Among Nations: The Struggle for Power and Peace*. 5th ed. New York: Alfred A. Knopf, 1973.

Mueller, Karl. "The Demise of Yugoslavia and the Destruction of Bosnia: Strategic Causes, Effects, and Responses." In *Deliberate Force: A Case Study in Effective Air Campaigning: Final Report of the Air University Balkans Air Campaign Study*, ed. Robert C. Owen, 1–36. Maxwell AFB, AL: Air University Press, 2000.

Nation, R. Craig. *War in the Balkans, 1991–2002*. Monograph. Carlisle, PA: Strategic Studies Institute, 2003.

The NATO Handbook: 50th Anniversary Edition. Brussels: Office of Information and Press, 1998.

Needless Deaths in the Gulf War: Civilian Casualties During the Air Campaign and Violations of the Laws of War. New York: Human Rights Watch, 1991.

Nef, John U. *War and Human Progress: An Essay on the Rise of Industrial Civilization*. Cambridge, MA: Harvard University Press, 1950.

Niebuhr, Reinhold. *Moral Man and Immoral Society: A Study in Ethics and Politics*. New York: Charles Scribner's Sons, 1932.

Orend, Brian. *War and International Justice: A Kantian Perspective*. Waterloo, Ont.: Wilfrid Laurier University Press, 2000.

Owen, Robert C. ed. *Deliberate Force: A Case Study in Effective Air Campaigning: Final Report of the Air University Balkans Air Campaign Study*. Maxwell AFB, AL: Air University Press, 2000.

Owen, Robert C. "Summary." In *Deliberate Force: A Case Study in Effective Air Campaigning: Final Report of the Air University Balkans Air Campaign Study*, ed. Robert C. Owen, 455–522. Maxwell AFB, AL: Air University Press, 2000.

O'Brien, William V. *The Conduct of Just and Limited War*. New York: Praeger, 1983.

O'Brien, William V., and John Langan, eds. *The Nuclear Dilemma and the Just War Tradition.* Lexington, MA: Lexington Books, 1986.

Pajić, Zoran. "Bosnia-Herzegovina: From Multiethnic Coexistence to 'Apartheid'... and Back." In *Yugoslavia the Former and the Future: Reflections by Scholars from the Region*, ed. Payam Akhavan and Robert Howse, 152–63. Washington, DC: Brookings Institution, 1995.

Palmer, Michael A. *On Course to Desert Storm: The United States Navy and the Persian Gulf.* Contributions to Naval History, vol. 5. Washington, DC: Naval Historical Center, 1992.

Palmer, R.R. (Robert Roswell), and Joel Colton. *A History of the Modern World.* 8th ed. New York: McGraw-Hill, 1995.

Pangle, Thomas L., and Peter J. Ahrensdorf. *Justice Among Nations: On the Moral Basis of Power and Peace*. Lawrence: University Press of Kansas, 1999.

Pape, Robert A. (Anthony) *Bombing to Win: Air Power and Coercion in War*. Ithaca, NY: Cornell University Press, 1996.

Payne, Keith B., and Karl I. Payne. *A Just Defense: The Use of Force, Nuclear Weapons, and Our Conscience*. Portland, OR: Multnomah, 1987.

Philosophy of Technology: The Technological Condition. An Anthology. ed. Robert C. Scharff and Val Dusek. Blackwell Philosophy Anthologies. Malden, MA: Blackwell Publishing, 2003.

The Political Writings of Thomas Jefferson: Representative Selections. ed. Edward Dumbauld. Reprinted. The American Heritage Series, vol. No. 9. New York: The Liberal Arts Press, 1956.

Posen, Barry R. "Inadvertent Nuclear War?: Escalation and NATO's Northern Flank." In *Strategy and Nuclear Deterrence: An International Security Reader*, ed. Steven E. Miller, 85–111. Princeton: Princeton University Press, 1984.

Postman, Neil. *Technopoly: The Surrender of Culture to Technology*. New York: Vintage Books, 1992.

Puhovski, Žarko. "Yugoslav Origins of the Post-Yugoslav Situation and the Bleak Prospects for Civil Society." In *Yugoslavia the Former and the Future: Reflections by Scholars from the Region*, ed. Payam Akhavan and Robert Howse, 121–38. Washington, DC: Brookings Institution, 1995.

Rawls, John [B]. *A Theory of Justice*. Cambridge, MA: Belknap Press, 1971.

——. *Political Liberalism*. New York: Columbia University Press, 1993.

Regan, Richard J. *Just War: Principles and Cases*. Washington, DC: Catholic University of America Press, 1996.

Reynolds, Richard T. *Heart of the Storm: The Genesis of the Air Campaign Against Iraq*. Maxwell AFB, AL: Air University Press, 1995.

Richelson, Jeffrey T. *America's Secret Eyes in Space: The U.S. Keyhole Spy Satellite Program*. New York: Harper & Row, 1970.

Robinson, Paul, ed. *Just War in Comparative Perspective*. Burlington, VT: Ashgate, 2003.

Russell, Bertrand. *Authority and the Individual: The Reith Lectures for 1948–9*. London: Unwin Brothers Ltd., 1949.

Sargent, Richard L. "Aircraft Used in Deliberate Force." In *Deliberate Force: A Case Study in Effective Air Campaigning: Final Report of the Air University Balkans Air Campaign Study*, ed. Robert C. Owen, 199–256. Maxwell AFB, AL: Air University Press, 2000.

——. "Deliberate Force Combat Air Assessment." In *Deliberate Force: A Case Study in Effective Air Campaigning: Final Report of the Air University Balkans Air Campaign Study*, ed. Robert C. Owen, 331–51. Maxwell AFB, AL: Air University Press, 2000.

——. "Deliberate Force Tactics." In *Deliberate Force: A Case Study in Effective Air Campaigning: Final Report of the Air University Balkans Air Campaign Study*, ed. Robert C. Owen, 297–330. Maxwell AFB, AL: Air University Press, 2000.

——. "Deliberate Force Targeting." In *Deliberate Force: A Case Study in Effective Air Campaigning: Final Report of the Air University Balkans Air Campaign Study*, ed. Robert C. Owen, 279–96. Maxwell AFB, AL: Air University Press, 2000.

——. "Weapons Used in Deliberate Force." In *Deliberate Force: A Case Study in Effective Air Campaigning: Final Report of the Air University Balkans Air Campaign Study*, ed. Robert C. Owen, 257–78. Maxwell AFB, AL: Air University Press, 2000.

Schwamb, Frank et al. *Desert Storm Reconstruction Report, Vol. II; Strike Warfare*. Center for Naval Analyses. Alexandria, VA, 1991.

Schaff, Philip. *The Confessions and Letters of St. Augustin, with a Sketch of His Life and Work*. New York: Christian Literature Publishing Co., 1886.

Schrijvers, Peter. *The Crash of Ruin: American Combat Soldiers in Europe During World War II*. London: Macmillan Press Ltd., 1998.

Schwarzkopf, H. Norman, and Peter Petre. *It Doesn't Take a Hero*. New York: Bantam, 1992.

Simonen, Katariina. *Operation Allied Force: A Case of Humanitarian Intervention*. Athena Papers. Garmisch-Partenkirchen: Partnership for Peace Consortium of Defense Academies and Security Studies Institutes, 2004.

Singer, J. David, and Melvin Small. *The Wages of War, 1816–1965: A Statistical Handbook*. New York: John Wiley and Sons, 1972.

Singer, P.W. (Peter Warren). *Wired for War: The Robotics Revolution and Conflict in the Twenty-first Century*. New York: Penguin Press, 2009.

Skolnikoff, Eugene B. *The Elusive Transformation: Science, Technology, and the Evolution of International Politics*. Princeton, NJ: Princeton University Press, 1993.

Slim, Hugo. *Killing Civilians: Method, Madness, and Morality in Warfare.* New York: Columbia University Press, 2008.

Stanik, Joseph T. *El Dorado Canyon: Reagan's Undeclared War with Qaddafi.* Annapolis: U.S. Naval Institute Press, 2003.

Strategy and Nuclear Deterrence: An International Security Reader. ed. by Steven E. Miller. Princeton: Princeton University Press, 1984.

Sutyagin, Igor. *The Tomahawk Missile System: Some Russian Views on Its Military and Political Significance*. Alexandria, VA: Center for Naval Analyses, 1998.

Tenner, Edward. *Why Things Bite Back: Technology and the Revenge of Unintended Consequences*. New York: Alfred A. Knopf, 1996.

Thomas, Ward. *The Ethics of Destruction: Norms and Force in International Relations*. Ithaca, NY: Cornell University Press, 2001.

Thucydides. *The Landmark Thucydides: A Comprehensive Guide to the Peloponnesian War*. ed. Robert B. Strassler. New York: The Free Press, 1996.

Tin, Bui. *From Enemy to Friend: A North Vietnamese Perspective on the War*. Translated by Nguyen Ngoc Bich. Annapolis: Naval Institute Press, 2002.

Toomay, John C. "Warning and Assessment Sensors." In *Managing Nuclear Operations*, ed. Ashton B Carter, John D. Steinbruner and Charles A. Zraket, 282–321. Washington, DC: Brookings Institution, 1987.

Trombley, Frank R. "The Arabs, the Byzantine State and the Islamic Law of War (*fiqh al-Jihad*) (7th–10th Centuries CE)." In *Just War in Comparative Perspective*, ed. Paul Robinson, 153–66. Burlington, VT: Ashgate, 2003.

Tse-Tung, Mao. *Selected Military Writings of Mao Tse-Tung*. Peking: Foreign Languages Press, 1967.

Tucker, Robert W. "1989 and All That." In *Sea-Changes: American Foreign Policy in a World Transformed*, ed. Nicholas X. Rizopoulos. New York: Council on Foreign Relations, 1990.

Twiss, Sumner B. "Religion and Human Rights: A Comparative Perspective." In *Explorations in Global Ethics: Comparative Religious Ethics and Interreligious Dialogue*, ed. Sumner B. Twiss and Bruce Grelle, 155–75. Boulder, CO: Westview Press, 2000.

Twiss, Sumner B., and Bruce Grelle, eds. *Explorations in Global Ethics: Comparative Religious Ethics and Interreligious Dialogue*. Boulder, CO: Westview Press, 2000.

Tzu, Sun. *The Art of War*. Translated by Samuel B. Griffith. Oxford: Oxford University Press, 1963.

The Use of Force: International Politics and Foreign Policy. ed. Robert J. Art and Kenneth N. Waltz. 2nd. ed. New York: University Press of America, 1983.

The Utilitarians. Garden City, New York: Dolphin Books, 1961.

Van Creveld, Martin. *Technology and War: From 2000 B.C. to the Present*. New York: The Free Press, 1991.

Vaux, Kenneth L. *Ethics and the Gulf War*. San Francisco: Westview Press, 1992.

Vojnić, Dragomar. "Disparity and Disintegration: The Economic Dimension of Yugoslavia's Demise." In *Yugoslavia the Former and the Future: Reflections by Scholars from the Region*, ed. Payam Akhavan and Robert Howse, 75–111. Washington, DC: Brookings Institution, 1995.

Walt, Stephen M. "International Relations: One World, Many Theories." In *Conflict and Cooperation: Evolving Theories of International Relations*, ed. Mark A. Genest, 29–38. Belmont, CA: Thomson Wadsworth, 2004.

Waltz, Kenneth N. *Theory of International Politics*. Reading, MA: Addison-Wesley, 1979.

Walzer, Michael. *Just and Unjust Wars: A Moral Argument with Historical Illustrations*. 3rd. ed. New York: Basic Books, 2000.

Warden, John A. III. *The Air Campaign: Planning for Combat*. Washington, DC: National Defense University Press, 1988.

——. *Planning to Win*. Air Power Studies Centre Papers, vol. P66. Fairbairn, Australia: Air Power Studies Centre, July, 1998.

Webster, Fr. Alexander F.C. "Justifiable War in Eastern Orthodox Christianity." In *Just War in Comparative Perspective*, ed. Paul Robinson, 40–61. Burlington, VT: Ashgate, 2003.

Werrell, Kenneth P. *Chasing the Silver Bullet: U.S. Air Force Weapons Development from Vietnam to Desert Storm*. Washington, DC: Smithsonian Books, 2003.

White, Paul K. *The Washington Institute for Near East Policy Military Research Papers*. Vol. 2, *Crises After the Storm: An Appraisal of U.S. Air Operations in Iraq Since the Persian Gulf War*. Washington, DC: The Washington Institute for Near East Policy, 1999.

Winnefeld, James A., and Dana J. Johnson. *Joint Air Operations: Pursuit of Unity in Command and Control, 1942–1991*. Annapolis: Naval Institute Press, 1993.

Woodward, Susan. *Balkan Tragedy: Chaos and Dissolution After the Cold War*. Washington, DC: Brookings Institution, 1995.

Zimmerman, Michael E. *Heidegger's Confrontation with Modernity: Technology, Politics, Art*. Bloomington: Indiana University Press, 1990.

Journals/Periodicals

Anderson, John Ward, and Omar Fekeiki. "Freedom, Yes, Iraqis Say, But at Great, Grave Cost." *Washington Post*, March 21 2006, 13.

Arkin, William. "Baghdad: The Urban Sanctuary in Desert Storm?" *Airpower Journal* XI, no. 1 (Spring 1997): 4–21.

——. "The Environmental Threat of Military Operations." In *U.S. Naval War College International Law Studies*. Vol. 69, 116–35. Newport: Naval War College Press, 1996.

——. "Tactical Bombing of Iraqi Forces Outstripped Value of Strategic Hits." *Aviation Week & Space Technology*, 27 January 1992, 62.

Bacevich, Andrew J. "Morality and High Technology." *National Interest* 45 (Fall 1996): 37–47.

Barry, John, and Evan Thomas. “Up in the Sky, An Unblinking Eye.” *Newsweek*, 9 June 2008, 24–9.

Bender, Bryan. “Tomahawk Achieves New Effects in Kosovo.” *Jane's Defence Weekly*, 19 July 2000.

——. “US Double Strike: ‘A New Ball Game.’” *Jane's Defence Weekly* 30.8 (26 August 1998): 8.

“Blitzkrieger.” *Time*, September 25 1939.

Bonsignore, Dr. Ezio. Letter to the editor. *Proceedings* 131.2 (February 2005): 17–18.

Brooks, David. “Rumsfeld's Blinkers.” *New York Times*, 16 March 2006, Late, A: 27.

Brooks, Harvey. “Technology, Evaluation, Purpose.” *Daedalus* 109.1 (Winter 1980): 65–81.

Budiansky et al., Stephen. “Air Power's Ultimate Test.” *US News and World Report*, 28 January 1991, 28–32, 39.

Christians, Clifford G. “Jacques Ellul's Conversions and Protestant Theology.” *Journal of Media and Religion* 5.3 (2006): 147–60.

Cook, Martin. “Michael Walzer's Concept of ‘Supreme Emergency.” *Journal of Military Ethics* 6.2 (2007): 138–51.

Cook, Nick. “War of Extremes.” *Jane's Defence Weekly* 32.1 (7 July 1999): 20–23.

Daum, Werner. “Universalism and the West: An Agenda for Understanding.” *Harvard International Review* (Cambridge, MA) 23.2 (Summer 2001): 19–23.

Dunham, Stephen. Letter to the editor. *Proceedings* 131.2 (February 2005): 18.

Dunlap, Charles J., Jr. “Technology: Recomplicating Moral Life for the Nation's Defenders.” *Parameters* XXIX, no. 3 (Autumn 1999): 24–53.

Ellul, Jacques. “Ideas of Technology: The Technological Order,” translated by John Wilkinson. *New Perspectives Quarterly* 17, no.3 (Summer 2000): 21–8.

Elsamer, Edgar. “Eglin's ADTC – Business End of the Air Force.” *Air Force Magazine* 55.8 (August 1972): 24–9.

——. “The Steady Evolution in Armaments.” *Air Force Magazine* 68.12 (December 1985): 72–81.

——. “The ‘Flyable’ Smart Bomb – Adding Another Dimension to Airpower.” *Air Force Magazine* 55.8 (August 1972): 30–35.

Erwin, Sandra I. “Washington Report: Air War Over Iraq Proves Need for Smaller, More Precise Ordnance.” *The Hook* 33.2 (Summer 2005): 68–9.

Evans, Michael. “Dark Victory.” *Proceedings* 125.9 (September 1999): 33–7.

Finn, Chris. “The Broader Implications of the Increasing Use of Precision Weapons.” *Air Power Review* 4.1 (Spring 2001): 34–56.

Fitzgerald, Mary C. “Marshal Ogarkov on Modern Theater Operations.” *Naval War College Review* 39.4 (Fall 1986): 6–25.

——. “Russia's New Military Doctrine.” *Naval War College Review* 46.2 (Spring 1993): 24–44.

——. “The Soviet Military and the New ‘Technological Operation’ in the Gulf.” *Naval War College Review* 44.4 (Fall 1991): 16–43.

Freedman, Lawrence. "Baghdad Stunned by Science." *The Independent*, January 18 1991, 21.

Friedman, Norman. "Is Modern War Too Precise?" *Proceedings* 130.12 (December 2004): 4–6.

——. Letter to the editor. *Proceedings* 131.3 (March 2005): 28.

Friedman, Thomas L. "The Big Question." *New York Times*, 3 March 2006, Late, A: 23.

Gardels, Nathan. "Post-Human History?" *New Perspectives Quarterly* 17.3 (Summer 2000): 2–3.

Gordon, Michael R. "Fateful Choice on Iraq Army Bypassed Debate." *New York Times*, 17 March 2008, 1.

Hanson, Victor Davis. "2,000 Dead, in Context." *New York Times*, 27 October 2005, Final, A: 31.

——. "Battles Change, Wars Don't." *Los Angeles Times*, 20 October 2005, California Metro, B: 11.

Hehir, J. Bryan. "Just War Theory in a Post-Cold War World." *Journal of Religious Ethics* 20.2 (Fall 1992): 237–58.

Hope, Samuel. "Homage to Jacques Ellul." *Arts Education Policy Review* 97.5 (May/June 1996): 38–9.

Hudak, Carl M. Letter to the editor. *Proceedings* 131.12 (January 2005): 19.

Hunsinger, George. "Iraq: Don't Go There." *Christian Century* 119.17 (August 2002): 10–12.

Iasiello, Rear Admiral Louis V. "Jus Post Bellum: The Moral Responsibilities of Victors in War." *Naval War College Review* 57.3/4 (Summer/Autumn 2004): 33–52.

Kaplan, Robert D. "Four-Star Generalists." *The Atlantic Monthly* 284.4 (October 1999): 18–19.

Kellogg, Davida E. "Jus Post Bellum: The Importance of War Crimes Trials." *Parameters* XXIII, no. 3 (Autumn 2002): 87–99.

Komarow, Steven. "Pentagon Boosts Number of U.S. Air Missions: Increased Use Influenced by Insurgencies." *USA Today*, 16 March 2006, 6.

Lancaster, John. "Advancing on High-Tech Weaponry." *Washington Post National Weekly Edition*, January 6–12 1992, 31.

Langston, Bud (Arthur N.), and Don Bringle. "Operation Praying Mantis: The Air View." *Proceedings* 115.5 (May 1989): 54–65.

Lennox, Duncan. "'Fox': The Results." *Jane's Defence Weekly* 31.2 (13 January 1999): 63.

Levin, Yuri. "Snow's Two Cultures – and Ours." *Public Interest* No. 153 (Fall 2003): 54–69.

Macdonald, Michael H. Review of *The Technological Bluff*, by Jacques Ellul (Translated by Geoffrey Bromiley). *The Christian Century* 108.22 (July 24–31 1991): 729–30.

Markow, David R. "The Russians and Their Nukes." *Air Force Magazine* 80, no.2 (February 1997): 40–43.

"Massacre of the Pure." *Time*, 28 April 1961, 54.

Meilinger, Phillip S. "Precision Aerospace Power, Discrimination, and the Future of War." *Royal Air Force Air Power Review* 4.3 (Autumn 2001): 14–26.

Molnar, Thomas. "Calling the Bluff." Review of *The Technological Bluff*, by Jacques Ellul (Translated by Geoffrey Bromiley). *National Review* 43.7 (April 29 1991): 53–5.

Mulrine, Anna. "Targeting the Enemy: Inside the Air Force's Control Center for Iraq and Afghanistan." *US News and World Report*, 9 June 2008, 26–8.

Orend, Brian. "Jus Post Bellum." *Journal of Social Philosophy* 31.1 (Spring 2000): 117–37.

Perkins III, J. (James) B. "Operation Praying Mantis: The Surface View." *Proceedings* 115.5 (May 1989): 66–70.

Perry, William J. "Desert Storm and Deterrence." *Foreign Affairs*, 70.4 (Fall 1991): 66–82.

Priest, Dana. "The Battle That Never Was: Did NATO's Threat to Send Ground Troops Bring Milosevic to the Table?" *Washington Post National Weekly Edition*, 27 September 1999, 6–7.

Reed, Fred. "Today's Weapons Overly High-Tech." *Washington Times*, 31 March 2006, C11.

"Report Puts Iraqi Dead at 1500." *Jane's Defence Weekly*, 13 March 1993, 5.

RGM/UGM-109B/C Tomahawk. Jane's Naval Weapons Systems, vol. Issue 26, April, 1998.

Richards, Evelyn. "From Smart Bombs to Brilliant Missiles." *Washington Post National Weekly Edition*, 11–17 March 1991, 34.

Risen, James. "To Bomb Sudan Plant, or Not: A Year Later, Debates Rankle." *New York Times* (New York), 27 October 1999, Late Edition – Final, A: 1.

Robertson, Horace B. Jr. "Contemporary International Law Relevant to Today's World?" *Naval War College Review* 45.1 (Summer 1992): 89–103.

Roy, Rustum. "Scientism and Technology as Religions." *Zygon* 40.4 (December 2005): 835–44.

Schmitt, Eric. "Pentagon Cites Some War Lessons." *New York Times*, 17 July 1991, A3.

Seigle, Greg. "Desert Fox Sparks Effectiveness Debate." *Jane's Defence Weekly* 6 January 1999: 2.

Sharkey, Noel. "Saying 'No!' to Lethal Autonomous Targeting." *Journal of Military Ethics* 9.4 (2010): 369–83.

Shea, Jamie. "Instant History." *NATO Review* 49.2 (Summer 2001): 21–3.

"Signals: US Navy Launches Tomahawk Strikes at Iraq." *Jane's Navy International* 101.8 (1 October 1996): 8.

Smith, Melden E., Jr. "The Strategic Bombing Debate: The Second World War and Vietnam." *Journal of Contemporary History* 12.1 (January 1977): 175–91.

Steinfels, Peter. "Beliefs; In the Heat of War, Confronting a Moral Principle of Keeping Civilians Immune from Direct Attack." *New York Times* (New York), 6 April 2002, Late, B: 6.

Struhl, Karsten J. "On Just War, Proportionality, and Bombing Civilians." *Radical Philosophy Review* 20.1 (1999): 18–20.

Tirpak, John A. "Kosovo Retrospective." *Air Force Magazine* 83.4 (April 2000): 28–33.

"U.S. Guided Bombs Alter Viet Air War." *Aviation Week & Space Technology*, 22 May 1972, 16–17.

Werrell, Kenneth P. "The Strategic Bombing of Germany in World War II: Costs and Accomplishments." *The Journal of American History* 73.3 (December 1986): 702–13.

Wilson, John. "Just an Ordinary Jacques." Review of *What I Believe*, by Jacques Ellul. *Christianity Today* 35.5 (April 29 1991): 52–3.

Zinn, Howard. "A Just Cause, Not a Just War." *The Progressive* 65.12 (December 2001): 16–19.

——. "What War Looks Like." *The Progressive* 66.10 (October 2002): 14–15.

Government Publications/Documents/Hearings

Bush, George (Herbert Walker). *National Security Directive 54: Responding to Iraqi Aggression in the Gulf*, 1991.

Charter of the United Nations, (1945).

Department of Defense. *Conduct of the Persian Gulf War: Final Report to Congress*. April 1992.

——. *Joint Publication (JP) 1-02: Department of Defense Dictionary of Military and Associated Terms*. Washington, DC: Government Printing Office, 12 April 2001 (As Amended Through 26 August 2008).

——. *Kosovo/Operation Allied Force After Action Report: Report to Congress*. 31 January 2000.

Keaney, Thomas A., and Eliot A. Cohen. *Gulf War Air Power Survey*. Washington, DC: Government Printing Office, 1993.

Metcalf, Roy E. "Guided TACAIR Weapons in Southeast Asia." Scientific Advisory Group Working Paper No. 4-72. Department of Defense, U.S. Pacific Command, 28 November 1972.

UN Universal Declaration of Human Rights, (1948).

U.S. Congress, Senate, Committee on Armed Services, Hearings before the Preparedness Investigating Subcommittee, *Air War Against North Vietnam*, 90th Cong., 1st sess., 22–3 August 1967.

U.S. Congress, Senate, Committee on Foreign Relations, *Bombing as a Policy Tool in Vietnam: Effectiveness*, 92nd Cong., 2nd sess., Study no. 5, 12 October 1972.

U.S. Congress, Senate, Committee on Foreign Relations, Subcommittee on U.S. Security Agreements and Commitments Abroad, Staff Report on *U.S. Air Operations in Cambodia: April 1973*, 93rd Cong., 1st sess., 27 April 1973.

U.S. Department of Defense. "Pentagon Papers." In *United States – Vietnam Relations 1945–1967*. Washington, DC: U.S. Government Printing Office, 1971.

Lectures/Speeches/Addresses

Bacon, Kenneth H., Assistant Secretary of Defense (Public Affairs). Press Conference. Washington, DC, 6 September 1996.

Bush, George H.W. (Herbert Walker), President. Address. White House, 8 August 1990.

——, President. *Aspen Institute Symposium*. Speech. Aspen, CO: Aspen Institute, 2 August 1990.

——, President. Press Conference. White House, 5 September 1990.

Bush, George W. "A Period of Consequences." Speech. The Citadel, 23 September 1999.

——, President. "President Bush Outlines Iraqi Threat." Cincinnati, OH: Cincinnati Museum Center, 7 October 2002.

Clinton, William Jefferson, President. Statement by the President. Martha's Vineyard, 20 August 1998.

Cook, Martin L. "Uncharted Waters." Lecture. U.S. Naval Academy, Annapolis, MD: William C. Stutt Lecture Series, 14 November 2005.

Hehir, J. Bryan. "New Challenges to an Ancient Ethic." Lecture. U.S. Naval Academy, Annapolis, MD: Spring Lecture Series, 9 February 2002.

Herring, George C. "In Cold Blood: LBJ's Conduct of Limited War in Vietnam." *No. 33*.

The Harmon Lecture Series in Military History. United States Air Force Academy, 17 October 1990.

Hoffman, Stanley. *The Political Ethics of International Relations*. Seventh Morgenthau Memorial Lecture on Ethics and Foreign Policy. New York: Carnegie Council on Ethics and International Relations, 1988.

Ignatieff, Michael. "Virtual War: Ethical Challenges." Lecture. U.S. Naval Academy, Annapolis, MD: Spring Lecture Series, 26 March 2001.

Orend, Brian. "Post-War Justice in Afghanistan and Iraq." Lecture. U.S. Naval Academy, Annapolis, MD: Spring Lecture Series, 19 March 2007.

Perry, William J, Secretary of Defense. Press Conference. Washington, DC, 3 September 1996.

Ralston, Joseph, Vice Chairman, JCS. Press Conference. Washington, DC, 4 September 1996.

Walzer, Michael. "The Triumph of Just War Theory (and the Dangers of Success)." Lecture. U.S. Naval Academy, Annapolis, MD: William C. Stutt Lecture Series, 18 November 2002.

Letters

Gregory, Bishop Wilton D., Washington, DC, United States Council of Catholic Bishops, to President George W. Bush, 13 September 2002.

Jefferson, Thomas, Poplar Forest, to Thomas Law. "The Moral Sense," 13 June 1814. University of Virginia Library, 1814161.

Unpublished Dissertation/Thesis/SPaper

Arkin, William M., Damian Durrant, and Marianne Cherni. "On Impact: Modern Warfare and the Environment: A Case Study of the Gulf War." Paper presented at the "Fifth Geneva" Convention on the Protection of the Environment in Time of Armed Conflict. London: A Roundtable Conference Jointly Organized by Centre for Defense Studies, King's College; Greenpeace International; London School of Economics, 3 June 1991.

Beale, Michael O. *Bombs Over Bosnia: The Role of Airpower in Bosnia-Herzegovina.* Thesis. Maxwell AFB, AL: Air University Press, 1997.

Christopher, Paul Pasquale. "Just War Theory: An Historical and Philosophical Analysis." Ph.D. diss., University of Massachusetts, 1990.

Gillespie, Paul G. "Precision Guided Munitions: Constructing a Bomb More Potent Than the A-Bomb." Ph.D. diss., History, Lehigh University, 2002.

Gray, Gary Grant. "Perceptions of Jacques Ellul's Educational Technique in a Modern Career-Focused M.B.A. Program." Ph.D. diss., Humanities Department, Salve Regina University, 2004.

Johnson, Michael W. "Just-War Theory and Future Warfare." M.Sc. Thesis, U.S. Army Command and General Staff College, 2000.

Jones, Laurence W. "The Persian Gulf War: A Case Study in Just War Theory." M.Sc. Thesis, Duke University, 1998.

Joynt, David M. "The Original Prophet and the Nuclear Revolution." Ph.D. diss., Princeton Theological Seminary, 1993.

Lohide, Kurtis D. *Desert Storm's Siren Song: Examining Revolution in Warfare.* Monograph. Fort Leavenworth, KS: School of Advanced Military Studies, 1994.

Lyddon, Peter. *The Fall and Rise of Coercive Diplomacy in the Balkans*. Monograph. Fort Leavenworth, KS: School of Advanced Military Studies, 1996.

Phinney, Todd R. *Airpower Versus Terrorism: Three Case Studies*. Thesis. Maxwell AFB, AL: School of Advanced Air and Space Studies, 2003.

Ream, Barry L. "Precision Guided Weapons Training and Employment." Research Report. Maxwell AFB, AL: Air War College, April 1988.

Tolbert, Julian H. *Crony Attack: Strategic Attack's Silver Bullet?* Thesis. Maxwell AFB, AL: Air University Press, 2006.

Walsh III, Thomas F. *Operation Allied Force: Setting a New Precedent for Humanitarian Intervention?* Thesis. Monterey, CA: Naval Post Graduate School, 2000.

Internet Resources

Appel, Lt Col Bernard, USAF. "Bombing Accuracy in a Combat Environment." *Air University Review* July/August 1975. 3 July 2008 http://www.airpower.maxwell.af.mil/airchronicles/aureview/1975/jul-aug/appel.html.

Aquinas, St. Thomas. "Of War (Four Articles)." In *Summa Theologica.* Christian Classics Ethereal Library. no date. Calvin College. 27 Sep 2006 <http://www.ccel.org/a/aquinas/summa/SS/SS040.html#SSQ40OUTP1>.

Boot, Max. "From Saigon to Desert Storm: How the U.S. Military Reinvented Itself After Vietnam." *American Heritage Magazine* 57.6 November/December 2006. 3 July 2008 http://www.americanheritage.com/articles/magazine/ah/2006/6/2006_6_28.shtml.

Bracken, Paul. "Technological Innovation and National Security." E-Note. 6 June 2008. Foreign Policy Research Institute. 18 July 2008 www.fpri.org/enotes/200806.bracken.innovationnationalsecurity.html.

Corn, Tony. "Clausewitz in Wonderland." *Policy Review* September 2006. Hoover Institution. 18 September 2006 http://www.policyreview.org/000/corn2.html.

Dempsey, Gary. "Destroying Serbia in Order to Save It." *CATO Institute* 8 June 1999. 17 December 2008 http://www.cato.org/pub_display.php?pub_id=5606.

Fowler, James A. "A Synopsis and Analysis of the Thought and Writings of Jacques Ellul." Article. 2000. 18 July 2006 http://www.christinyou.net/pages/ellul.html.

McIntyre, Jamie. "'Moderate to Severe' Damage Seen at Suspected Bin Laden Camps." *CNN.Com* 13 January 1999. 29 December 2008 http://www.cnn.com/WORLD/asiapcf/9901/13/afghan.damage.photos.

Moseley, Alexander. "Just War Theory." In *The Internet Encyclopedia of Philosophy*. 2006. 10 March 2006 http://www.iep.utm.edu/j/justwar.htm.

Raffaele, Paul. "Out of Time." *Smithsonian Magazine* April 2005: Travel. 25 September 2007 http://www.smithsonianmagazine.com/issues/2005/april/time.php.

Raymond, Brother John. *The Just War Theory*. n.d. 26 Sep 2006 http://www.monksofadoration.org/justwar.html.

Schaefer, David Lewis. "What Did Gandhi Do? One Sided Pacifist." *National Review Online* 23 April 2003. 19 September 2006 http://www.nationalreview.com/comment/comment-schaefer042803.asp.

Index

For Product Safety Concerns and Information please contact our EU
representative GPSR@taylorandfrancis.com
Taylor & Francis Verlag GmbH, Kaufingerstraße 24, 80331 München, Germany

www.ingramcontent.com/pod-product-compliance
Ingram Content Group UK Ltd.
Pitfield, Milton Keynes, MK11 3LW, UK
UKHW010959180926
475701UK00004B/449

* 9 7 8 1 1 3 8 1 1 5 6 9 9 *